Studies in Computational Intelligence

Volume 649

Series editor

Janusz Kacprzyk, Polish Academy of Sciences, Warsaw, Poland
e-mail: kacprzyk@ibspan.waw.pl

About this Series

The series "Studies in Computational Intelligence" (SCI) publishes new developments and advances in the various areas of computational intelligence—quickly and with a high quality. The intent is to cover the theory, applications, and design methods of computational intelligence, as embedded in the fields of engineering, computer science, physics and life sciences, as well as the methodologies behind them. The series contains monographs, lecture notes and edited volumes in computational intelligence spanning the areas of neural networks, connectionist systems, genetic algorithms, evolutionary computation, artificial intelligence, cellular automata, self-organizing systems, soft computing, fuzzy systems, and hybrid intelligent systems. Of particular value to both the contributors and the readership are the short publication timeframe and the worldwide distribution, which enable both wide and rapid dissemination of research output.

More information about this series at http://www.springer.com/series/7092

George A. Anastassiou · Ioannis K. Argyros

Intelligent Numerical Methods II: Applications to Multivariate Fractional Calculus

George A. Anastassiou
Department of Mathematical Sciences
The University of Memphis
Memphis, TN
USA

Ioannis K. Argyros
Department of Mathematical Sciences
Cameron University
Lawton, OK
USA

ISSN 1860-949X ISSN 1860-9503 (electronic)
Studies in Computational Intelligence
ISBN 978-3-319-81558-9 ISBN 978-3-319-33606-0 (eBook)
DOI 10.1007/978-3-319-33606-0

Printed on acid-free paper

This Springer imprint is published by Springer Nature
The registered company is Springer International Publishing AG Switzerland

To our wives: Diana and Koula

Preface

This is a supplementary, complementary, and companion booklet monograph to the recently published monograph, by the same authors: "Intelligent Numerical Methods: Applications to Fractional Calculus", Studies in Computational Intelligence 624, Springer Heidelberg New York, 2016. It is the analog of the last one, regarding applications of Newton-like and other similar methods for solving multivariate equations, which involve Caputo-type fractional mixed partial derivatives and multivariate fractional Riemann–Liouville integral operators. These are studied for the first time in the literature, and chapters are self-contained and can be read independently. This booklet monograph is suitable to be used in related graduate classes and research projects. We exhibit the maximum of these numerical methods at the fractional multivariate level.

The list of presented topics follows:

A fixed point convergence theorem with applications in left multivariate fractional calculus.

Fixed point schemes with applications in right multivariate fractional calculus.

Results on the semilocal convergence of iterative methods with applications to k-multivariate fractional calculus.

Newton-like methods and their applications in multivariate fractional calculus.

Implicit iterative methods for solving equations with applications in multivariate calculus.

On the monotone convergence of general iterative methods with applications in fractional calculus.

Extending the applicability of the local and semilocal convergence of Newton's method.

On left multidimensional Riemann–Liouville fractional integral.

On right multidimensional Riemann–Liouville fractional integral.

For the last two topics, see Chaps. 8–9. These were studied on the sole purpose to support the fractional essential parts of Chaps. 1–3.

An extensive list of references is given per chapter.

The book's results are expected to find applications in many areas of applied mathematics, stochastics, computer science, and engineering. As such this short monograph is suitable for researchers, graduate students, and seminars of the above subjects, also to be in all science and engineering libraries.

The preparation of the book took place during 2015–2016 in Memphis, Tennessee and Lawton, Oklahoma, USA.

We would like to thank Prof. Alina Lupas of University of Oradea, Romania, for checking and reading the manuscript.

Memphis
Lawton
March 2016

George A. Anastassiou
Ioannis K. Argyros

Contents

About the Authors

George A. Anastassiou was born in Athens, Greece in 1952. He received his B.Sc. degree in Mathematics from Athens University, Greece in 1975. He received his Diploma in Operations Research from Southampton University, UK in 1976. He also received his MA in Mathematics from University of Rochester, USA in 1981. He was awarded his Ph.D. in Mathematics from University of Rochester, USA in 1984. During 1984–1986, he served as Visiting Assistant Professor at the University of Rhode Island, USA. Since 1986, he has been a faculty member at the University of Memphis, USA. He is currently a full Professor of Mathematics since 1994. His research area is "Computational Analysis" in a very broad sense. He has published over 400 research articles in international mathematical journals and over 27 monographs, proceedings, and textbooks in well-known publishing houses. Several awards have been awarded to George Anastassiou. In 2007 he received the Honorary Doctoral Degree from University of Oradea, Romania. He is associate editor in over 60 international mathematical journals and editor in-chief in three journals, most notably in the well-known "Journal of Computational Analysis and Applications".

Ioannis K. Argyros was born in Athens, Greece in 1956. He received his B.Sc. degree in Mathematics from Athens University, Greece in 1979. He also received his MA in Mathematics from University of Georgia, USA in 1983. He was awarded his Ph.D. in Mathematics from University of Georgia, USA in 1984. During 1984–1986, he served as Visiting Assistant Professor at the University of Iowa, USA. During 1986–1990, he also served as Assistant Professor at the New Mexico State University, USA. Since 1990, he has been a faculty member at Cameron University, USA. He is currently a full Professor of Mathematics since 1994. His research area is "Computational Mathematics" in a very broad sense. He has published over 850 research articles in national and international mathematical journals and over 25 monographs, proceedings, and textbooks in well-known publishing houses. Several recognitions have been awarded to Ioannis K. Argyros. In 2001 he received the "Distinguished Research Award" from the Southwest Oklahoma Advanced Technology Association.

He is associate editor in over 50 international mathematical journals, most notably in the well-known "Applied Mathematics and Computation" and editor in-chief in one journal.

Chapter 1
Fixed Point Results and Their Applications in Left Multivariate Fractional Calculus

A fixed point theorem is given under general conditions on the operators involved in a Banach space setting. The results find applications in left multivariate fractional calculus. It follows [8].

1.1 Introduction

Numerous problems can be formulated as an equation like

$$R(x) = 0, \tag{1.1.1}$$

where R is a continuous operator defined on a subset Ω of a Banach space B_1 with values in a Banach space B_2 using Mathematical Modelling [1, 7, 12, 13, 17, 19]. The solutions denoted by x^* can be found in explicit form only in special cases. That is why most solution methods for these equations are usually iterative. Let $\mathcal{L}(B_1, B_2)$ denote the space of bounded linear operators from B_1 into B_2. Let also $A(\cdot) : \Omega \to \mathcal{L}(B_1, B_1)$ be a continuous operator. Set

$$F = LR, \tag{1.1.2}$$

where $L \in \mathcal{L}(B_2, B_1)$. We shall approximate x^* using a sequence $\{x_n\}$ generated by the fixed point scheme:

$$\begin{aligned} &x_{n+1} := x_n + z_n, \quad A(x_n) z_n + F(x_n) = 0 \\ &\Leftrightarrow z_n = Q(z_n) := (I - A(x_n)) z_n - F(x_n), \end{aligned} \tag{1.1.3}$$

where $x_0 \in \Omega$. The sequence $\{x_n\}$ defined by

$$x_{n+1} = Q(x_n) = Q^{(n+1)}(x_0) \tag{1.1.4}$$

G.A. Anastassiou and I.K. Argyros, *Intelligent Numerical Methods II: Applications to Multivariate Fractional Calculus*, Studies in Computational Intelligence 649, DOI 10.1007/978-3-319-33606-0_1

exists. In case of convergence we write:

$$Q^{\infty}(x_0) := \lim_{n\to\infty}\left(Q^n(x_0)\right) = \lim_{n\to\infty} x_n. \qquad (1.1.5)$$

Many methods in the literature can be considered special cases of method (1.1.3). We can choose A to be: $A(x) = F'(x)$ (Newton's method), $A(x) = F'(x_0)$ (Modified Newton's method), $A(x) = [x, g(x); F]$, $g : \Omega \to B_1$ (Steffensen's method). Many other choices for A can be found in [1–21] and the references there in. Therefore, it is important to study the convergence of method (1.1.3) under generalized conditions. In particular, we present the semi-local convergence of method (1.1.3) using only continuity assumptions on operator F and for a so general operator A as to allow applications to left multivariate fractional calculus and other areas.

The rest of the chapter is organized as follows: Sect. 1.2 contains the semi-local convergence of method (1.1.3). In the concluding Sect. 1.3, we suggest some applications to left multivariate fractional calculus.

1.2 Convergence

Let $B(w,\xi)$, $\overline{B}(w,\xi)$ stand, respectively for the open and closed balls in B_1 with center $w \in B_1$ and of radius $\xi > 0$.

We present the semi-local convergence of method (1.1.3) in this section.

Theorem 1.1 *Let $F : \Omega \subset B_1 \to B_2$, $A(\cdot) : \Omega \to \mathcal{L}(B_1, B_1)$ and $x_0 \in \Omega$ be as defined in the Introduction. Suppose: there exist $\delta_0 \in (0,1)$, $\delta_1 \in (0,1)$, $\eta \geq 0$ such that for each $x, y \in \Omega$*

$$\delta := \delta_0 + \delta_1 < 1, \qquad (1.2.1)$$

$$\|F(x_0)\| \leq \eta, \qquad (1.2.2)$$

$$\|I - A(x)\| \leq \delta_0, \qquad (1.2.3)$$

$$\|F(y) - F(x) - A(x)(y - x)\| \leq \delta_1 \|y - x\| \qquad (1.2.4)$$

and

$$\overline{B}(x_0, \delta) \subseteq \Omega, \qquad (1.2.5)$$

where

$$\rho = \frac{\eta}{1-\delta}. \qquad (1.2.6)$$

Then, sequence $\{x_n\}$ generated for $x_0 \in \Omega$ by

$$x_{n+1} = x_n + Q^{\infty}(0), \quad Q_n(z) := (I - A(x_n))z - F(x_n) \qquad (1.2.7)$$

is well defined in $\overline{B}(x_0, \rho)$, *remains in* $\overline{B}(x_0, \rho)$ *for each* $n = 0, 1, 2, \ldots$ *and converges to* x^* *which is the only solution of equation* $F(x) = 0$ *in* $\overline{B}(x_0, \rho)$. *Moreover, an apriori error estimate is given by the sequence* $\{\rho_n\}$ *defined by*

$$\rho_0 := \rho, \quad \rho_n = T_n^{\infty}(0), \quad T_n(t) = \delta_0 + \delta_1 \rho_{n-1} \tag{1.2.8}$$

for each $n = 1, 2, \ldots$ *and satisfying*

$$\lim_{n\to\infty} \rho_n = 0. \tag{1.2.9}$$

Furthermore, an aposteriori error estimate is given by the sequence $\{\sigma_n\}$ *defined by*

$$\sigma_n := H_n^{\infty}(0), \quad H_n(t) = \delta t + \delta_1 p_{n-1}, \tag{1.2.10}$$

$$q_n := \|x_n - x_0\| \leq \rho - \rho_n \leq \rho, \tag{1.2.11}$$

where

$$p_{n-1} := \|x_n - x_{n-1}\| \quad \textit{for each} \quad n = 1, 2, \ldots \tag{1.2.12}$$

Proof We shall show using mathematical induction the following assertion is true:
(A_n) $x_n \in X$ and $\rho_n \geq 0$ are well defined and such that

$$\rho_n + p_{n-1} \leq \rho_{n-1}. \tag{1.2.13}$$

By the definition of ρ, (1.2.3)–(1.2.6) we have that there exists $r \leq \rho$ (Lemma 1.4 [7, pp. 3]) such that

$$\delta_0 \tau + \|F(x_0)\| = r$$

and

$$\delta_0^k r \leq \delta_0^k \rho \to 0 \text{ as } k \to \infty.$$

That is (Lemma 1.5 [7, pp. 4]) x_1 is well defined and $p_0 \leq r$.

We need the estimate:

$$T_1(\rho - r) = \delta_0(\rho - r) + \delta_1 \rho_0 =$$

$$\delta_0 \rho - \delta_0 r + \delta_1 \rho = G_0(\rho) - r = \rho - r.$$

That is (Lemma 1.4 [7, pp. 3]) ρ_1 exists and satisfies

$$\rho_1 + p_0 \leq \rho - r + r = \rho = \rho_0.$$

Hence (I_0) is true. Suppose that for each $k = 1, 2, ..., n,$ assertion (I_k) is true. We must show: x_{k+1} exists and find a bound r for p_k. Indeed, we have in turn that

$$\begin{aligned}\delta_0\rho_k + \delta_1\left(\rho_{k-1} - \rho_k\right) &= \delta_0\rho_k + \delta_1\rho_{k-1} - \delta_1\rho_k \\ &= T_k\left(\rho_k\right) - \delta_1\rho_k \le \rho_k.\end{aligned}$$

That is there exists $r \le \rho_k$ such that

$$r = \delta_0 r + \delta_1\left(\rho_{k-1} - \rho_k\right) \quad \text{and} \quad \left(\delta_0 + \delta_1\right)^i r \to 0 \tag{1.2.14}$$

as $i \to \infty$.

The induction hypothesis gives that

$$q_k \le \sum_{m=0}^{k-1} p_m \le \sum_{m=0}^{k-1}\left(\rho_m - \rho_{m+1}\right) = \rho - \rho_k \le \rho,$$

so $x_k \in \overline{B}\left(x_0, \rho\right) \subseteq \Omega$ and x_1 satisfies $\|I - A\left(x_1\right)\| \le \delta_0$ (by (1.2.3)).

Using the induction hypothesis, (1.1.3) and (1.2.4), we get

$$\begin{aligned}\|F\left(x_k\right)\| &= \|F\left(x_k\right) - F\left(x_{k-1}\right) - A\left(x_{k-1}\right)\left(x_k - x_{k-1}\right)\| \\ &\le \delta_1 p_{k-1} \le \delta_1\left(\rho_{k-1} - \rho_k\right)\end{aligned} \tag{1.2.15}$$

leading together with (1.2.14) to:

$$\delta_0 r + \|F\left(x_k\right)\| \le r,$$

which implies x_{k+1} exists and $p_k \le r \le \rho_k$. It follows from the definition of ρ_{k+1} that

$$T_{k+1}\left(\rho_k - r\right) = T_k\left(\rho_k\right) - r = \rho_k - r,$$

so ρ_{k+1} exists and satisfies

$$\rho_{k+1} + p_k \le \rho_k - r + r = \rho_k$$

so the induction for (I_n) is completed.

Let $j \ge k$. Then, we obtain in turn that

$$\left\|x_{j+k} - x_k\right\| \le \sum_{i=k}^{j} p_i \le \sum_{i=k}^{j}\left(\rho_j - \rho_{j+1}\right) = \rho_k - \rho_{j+k} \le \rho_k. \tag{1.2.16}$$

We also have using induction that

$$\rho_{k+1} = T_{k+1}\left(\rho_{k+1}\right) \le T_{k+1}\left(\rho_k\right) \le \delta\rho_k \le ... \le \delta^{k+1}\rho. \tag{1.2.17}$$

Hence, by (1.2.1) and (1.2.17) $\lim_{k\to\infty} \rho_k = 0$, so $\{x_k\}$ is a complete sequence in a Banach space X and as such it converges to some x^*. By letting $j \to \infty$ in (1.2.16), we conclude that $x^* \in \overline{B}(x_k, \rho_k)$. Moreover, by letting $k \to \infty$ in (1.2.15) and using the continuity of F we get that $F(x^*) = 0$. Notice that

$$H_k(\rho_k) \leq T_k(\rho_k) \leq \rho_k,$$

so the apriori bound exists. That is σ_k is smaller in general than ρ_k. Clearly, the conditions of the theorem are satisfied for x_k replacing x_0 (by (1.2.16)). Hence, by (1.2.8) $x^* \in \overline{B}(x_n, \sigma_n)$, which completes the proof for the aposteriori bound. ■

Remark 1.2 (a) It follows from the proof of Theorem 1.1 that the conclusions hold, if $A(\cdot)$ is replaced by a more general continuous operator $A : \Omega \to B_1$.

(b) In the next section some applications are suggested for special choices of the "A" operators with $\gamma_0 := \delta_0$ and $\gamma_1 := \delta_1$.

1.3 Applications to Left Multivariate Fractional Calculus

Our presented earlier semi-local convergence results, see Theorem 1.1, apply in the next two multivariate fractional settings given that the following inequalities are fulfilled:

$$\|1 - A(x)\|_{\infty} \leq \gamma_0 \in (0, 1), \tag{1.3.1}$$

and

$$\left\|(F(y) - F(x))\overrightarrow{i} - A(x)(y - x)\right\| \leq \gamma_1 \|y - x\|, \tag{1.3.2}$$

where $\gamma_0, \gamma_1 \in (0, 1)$, furthermore

$$\gamma = \gamma_0 + \gamma_1 \in (0, 1), \tag{1.3.3}$$

for all $x, y \in \prod_{i=1}^{k} [a_i^*, b_i^*]$, where $a_i < a_i^* < b_i^* < b_i$, $i = 1, ..., k$.

Above $\overrightarrow{i}$ is the unit vector in $\mathbb{R}^k$, $k \in \mathbb{N}$, $\left\|\overrightarrow{i}\right\| = 1$, and $\|\cdot\|$ is a norm in $\mathbb{R}^k$.

The specific functions $A(x)$, $F(x)$ will be described next.

(I) Consider the left multidimensional Riemann–Liouville fractional integral of order $\alpha = (\alpha_1, ..., \alpha_k)$ $(\alpha_i > 0, i = 1, ..., k)$:

$$\left(I_{a+}^{\alpha} f\right)(x) = \frac{1}{\prod_{i=1}^{k} \Gamma(\alpha_i)} \int_{a_1}^{x_1} ... \int_{a_k}^{x_k} \prod_{i=1}^{k} (x_i - t_i)^{\alpha_i - 1} f(t_1, ..., t_k)\, dt_1 ... dt_k, \tag{1.3.4}$$

where Γ is the gamma function, $f \in L_\infty\left(\prod_{i=1}^{k}[a_i, b_i]\right)$, $a = (a_1, ..., a_k)$, and $x = (x_1, ..., x_k) \in \prod_{i=1}^{k}[a_i, b_i]$.

By [6], we get that $\left(I_{a+}^{\alpha} f\right)$ is a continuous function on $\prod_{i=1}^{k}[a_i, b_i]$. Furthermore by [6] we get that I_{a+}^{α} is a bounded linear operator, which is a positive operator, plus that $\left(I_{a+}^{\alpha} f\right)(a) = 0$.

In particular, $\left(I_{a+}^{\alpha} f\right)$ is continuous on $\prod_{i=1}^{k}\left[a_i^*, b_i^*\right]$.

Thus there exist $x_1, x_2 \in \prod_{i=1}^{k}\left[a_i^*, b_i^*\right]$ such that

$$\begin{aligned} \left(I_{a+}^{\alpha} f\right)(x_1) &= \min \left(I_{a+}^{\alpha} f\right)(x), \\ \left(I_{a+}^{\alpha} f\right)(x_2) &= \max \left(I_{a+}^{\alpha} f\right)(x), \end{aligned} \tag{1.3.5}$$

over all $x \in \prod_{i=1}^{k}\left[a_i^*, b_i^*\right]$.

We assume that

$$\left(I_{a+}^{\alpha} f\right)(x_1) > 0. \tag{1.3.6}$$

Hence

$$\left\| I_{a+}^{\alpha} f \right\|_{\infty, \prod_{i=1}^{k}\left[a_i^*, b_i^*\right]} = \left(I_{a+}^{\alpha} f\right)(x_2) > 0. \tag{1.3.7}$$

Here, we define

$$Jf(x) = mf(x), \quad 0 < m < \frac{1}{2}, \tag{1.3.8}$$

for any $x \in \prod_{i=1}^{k}\left[a_i^*, b_i^*\right]$.

Therefore the equation

$$Jf(x) = 0, \quad x \in \prod_{i=1}^{k}\left[a_i^*, b_i^*\right], \tag{1.3.9}$$

has the same solutions as the equation

$$F(x) := \frac{Jf(x)}{2\left(I_{a+}^{\alpha} f\right)(x_2)} = 0, \quad x \in \prod_{i=1}^{k}\left[a_i^*, b_i^*\right]. \tag{1.3.10}$$

Notice that

$$I_{a+}^{\alpha}\left(\frac{f}{2\left(I_{a+}^{\alpha}f\right)(x_2)}\right)(x)=\frac{\left(I_{a+}^{\alpha}f\right)(x)}{2\left(I_{a+}^{\alpha}f\right)(x_2)}\leq\frac{1}{2}<1,\quad x\in\prod_{i=1}^{k}\left[a_i^*,b_i^*\right].\tag{1.3.11}$$

Call

$$A(x):=\frac{\left(I_{a+}^{\alpha}f\right)(x)}{2\left(I_{a+}^{\alpha}f\right)(x_2)},\quad\forall\,x\in\prod_{i=1}^{k}\left[a_i^*,b_i^*\right].\tag{1.3.12}$$

We notice that

$$0<\frac{\left(I_{a+}^{\alpha}f\right)(x_1)}{2\left(I_{a+}^{\alpha}f\right)(x_2)}\leq A(x)\leq\frac{1}{2},\quad\forall\,x\in\prod_{i=1}^{k}\left[a_i^*,b_i^*\right].\tag{1.3.13}$$

Hence, the first condition (1.3.1) is fulfilled by

$$|1-A(x)|=1-A(x)\leq 1-\frac{\left(I_{a+}^{\alpha}f\right)(x_1)}{2\left(I_{a+}^{\alpha}f\right)(x_2)}=:\gamma_0,\quad\forall\,x\in\prod_{i=1}^{k}\left[a_i^*,b_i^*\right].\tag{1.3.14}$$

Hence, $\|1-A(x)\|_{\infty}\leq\gamma_0$, where $\|\cdot\|_{\infty}$ is over $\prod_{i=1}^{k}\left[a_i^*,b_i^*\right]$. Clearly $\gamma_0\in(0,1)$.

Next, we assume that $\frac{f(x)}{2\left(I_{a+}^{\alpha}f\right)(x_2)}$ is a contraction, that is

$$\left|\frac{f(x)}{2\left(I_{a+}^{\alpha}f\right)(x_2)}-\frac{f(y)}{2\left(I_{a+}^{\alpha}f\right)(x_2)}\right|\leq\theta\,\|x-y\|,\quad\text{all}\,x,y\in\prod_{i=1}^{k}\left[a_i^*,b_i^*\right],\tag{1.3.15}$$

$0<\theta<1$.

Hence

$$\left|\frac{mf(x)}{2\left(I_{a+}^{\alpha}f\right)(x_2)}-\frac{mf(y)}{2\left(I_{a+}^{\alpha}f\right)(x_2)}\right|\leq m\theta\,\|x-y\|\leq\frac{\theta}{2}\,\|x-y\|,\tag{1.3.16}$$

all $x,y\in\prod_{i=1}^{k}\left[a_i^*,b_i^*\right]$.

Set $\lambda=\frac{\theta}{2}$, it is $0<\lambda<\frac{1}{2}$. We have that

$$|F(x)-F(y)|\leq\lambda\,\|x-y\|,\tag{1.3.17}$$

all $x,y\in\prod_{i=1}^{k}\left[a_i^*,b_i^*\right]$.

Equivalently we have

$$|Jf(x) - Jf(y)| \leq 2\lambda \left(I_{a+}^{\alpha} f\right)(x_2) \|x - y\|, \quad \text{all } x, y \in \prod_{i=1}^{k} \left[a_i^*, b_i^*\right]. \quad (1.3.18)$$

We observe that

$$\left\| (F(y) - F(x)) \overrightarrow{i} - A(x)(y - x) \right\| \leq$$

$$|F(y) - F(x)| + |A(x)| \|y - x\| \leq \quad (1.3.19)$$

$$\lambda \|y - x\| + |A(x)| \|y - x\| = (\lambda + |A(x)|) \|y - x\| =: (\psi_1)$$

$\forall\, x, y \in \prod_{i=1}^{k} \left[a_i^*, b_i^*\right]$.

By [6], we have that

$$\left|\left(I_{a+}^{\alpha} f\right)(x)\right| \leq \left(\prod_{i=1}^{k} \frac{(b_i - a_i)^{\alpha_i}}{\Gamma(\alpha_i + 1)}\right) \|f\|_{\infty}, \quad (1.3.20)$$

$\forall\, x \in \prod_{i=1}^{k} \left[a_i^*, b_i^*\right]$, where $\|\cdot\|_{\infty}$ now is over $\prod_{i=1}^{k} [a_i, b_i]$.

Hence

$$|A(x)| = \frac{\left|\left(I_{a+}^{\alpha} f\right)(x)\right|}{2\left(I_{a+}^{\alpha} f\right)(x_2)} \leq \frac{1}{2\left(I_{a+}^{\alpha} f\right)(x_2)} \left(\prod_{i=1}^{k} \frac{(b_i - a_i)^{\alpha_i}}{\Gamma(\alpha_i + 1)}\right) \|f\|_{\infty} < \infty, \quad (1.3.21)$$

$\forall\, x \in \prod_{i=1}^{k} \left[a_i^*, b_i^*\right]$.

Therefore we get

$$(\psi_1) \leq \left(\lambda + \frac{1}{2\left(I_{a+}^{\alpha} f\right)(x_2)} \left(\prod_{i=1}^{k} \frac{(b_i - a_i)^{\alpha_i}}{\Gamma(\alpha_i + 1)}\right) \|f\|_{\infty}\right) \|y - x\|, \quad (1.3.22)$$

$\forall\, x, y \in \prod_{i=1}^{k} \left[a_i^*, b_i^*\right]$.

Call

$$0 < \gamma_1 := \lambda + \frac{1}{2\left(I_{a+}^{\alpha} f\right)(x_2)} \left(\prod_{i=1}^{k} \frac{(b_i - a_i)^{\alpha_i}}{\Gamma(\alpha_i + 1)}\right) \|f\|_{\infty}, \quad (1.3.23)$$

and by choosing $(b_i - a_i)$ small enough, $i = 1, \ldots, k$, we can make $\gamma_1 \in (0, 1)$, fulfilling (1.3.2).

Next, we call and we need that

$$0<\gamma:=\gamma_0+\gamma_1=\left(1-\frac{\left(I_{a+}^{\alpha}f\right)(x_1)}{2\left(I_{a+}^{\alpha}f\right)(x_2)}\right)+$$

$$\left(\lambda+\frac{1}{2\left(I_{a+}^{\alpha}f\right)(x_2)}\left(\prod_{i=1}^{k}\frac{(b_i-a_i)^{\alpha_i}}{\Gamma(\alpha_i+1)}\right)\|f\|_{\infty}\right)<1, \tag{1.3.24}$$

equivalently,

$$\lambda+\frac{1}{2\left(I_{a+}^{\alpha}f\right)(x_2)}\left(\prod_{i=1}^{k}\frac{(b_i-a_i)^{\alpha_i}}{\Gamma(\alpha_i+1)}\right)\|f\|_{\infty}<\frac{\left(I_{a+}^{\alpha}f\right)(x_1)}{2\left(I_{a+}^{\alpha}f\right)(x_2)}, \tag{1.3.25}$$

equivalently,

$$2\lambda\left(I_{a+}^{\alpha}f\right)(x_2)+\left(\prod_{i=1}^{k}\frac{(b_i-a_i)^{\alpha_i}}{\Gamma(\alpha_i+1)}\right)\|f\|_{\infty}<\left(I_{a+}^{\alpha}f\right)(x_1), \tag{1.3.26}$$

which is possible for small λ and small (b_i-a_i), all $i=1,...,k$. That is $\gamma\in(0,1)$, fulfilling (1.3.3). So our numerical method converges and solves (1.3.9).

(II) Let $\alpha=(\alpha_1,...,\alpha_k)$, $\alpha_i>0$, $m_i=\lceil\alpha_i\rceil$ ($\lceil\cdot\rceil$ ceiling function), $\alpha_i\notin\mathbb{N}$, $i=1,...,k\in\mathbb{N}$, and $G\in C^{\sum_{i=1}^{k}m_i-1}\left(\prod_{i=1}^{k}[a_i,b_i]\right)$, such that

$$0\neq\frac{\partial^{\sum_{i=1}^{k}m_i}G}{\partial x_1^{m_1}...\partial x_k^{m_k}}\in L_{\infty}\left(\prod_{i=1}^{k}[a_i,b_i]\right).$$

Here we consider the multivariate left Caputo type fractional mixed partial derivative of order α:

$$D_{*a}^{\alpha}G(x)=\frac{1}{\prod_{i=1}^{k}\Gamma(m_i-\alpha_i)}\int_{a_1}^{x_1}...\int_{a_k}^{x_k}\prod_{i=1}^{k}(x_i-t_i)^{m_i-\alpha_i-1}\cdot \tag{1.3.27}$$

$$\frac{\partial^{\sum_{i=1}^{k}m_i}G(t_1,...,t_k)}{\partial t_1^{m_1}...\partial t_k^{m_k}}dt_1...dt_k,$$

where again Γ is the gamma function, $a=(a_1,...,a_k)$, $\forall\ x=(x_1,...,x_k)\in\prod_{i=1}^{k}[a_i,b_i]$. Notice here that $m_i-\alpha_i>0$, $i=1,...,k$.

By [6], we get that $D_{*a}^{\alpha}G$ is a continuous function on $\prod_{i=1}^{k}[a_i, b_i]$, and it holds that $D_{*a}^{\alpha}G(a) = 0$.

In particular $D_{*a}^{\alpha}G$ is continuous on $\prod_{i=1}^{k}[a_i^*, b_i^*]$, where $a_i < a_i^* < b_i^* < b_i$, $i = 1, ..., k$.

Therefore there exist $x_1, x_2 \in \prod_{i=1}^{k}[a_i^*, b_i^*]$ such that

$$\begin{aligned} \left(D_{*a}^{\alpha}G\right)(x_1) &= \min \left(D_{*a}^{\alpha}G\right)(x), \\ \left(D_{*a}^{\alpha}G\right)(x_2) &= \max \left(D_{*a}^{\alpha}G\right)(x), \end{aligned} \tag{1.3.28}$$

over all $x \in \prod_{i=1}^{k}[a_i^*, b_i^*]$.

We assume that

$$\left(D_{*a}^{\alpha}G\right)(x_1) > 0. \tag{1.3.29}$$

Hence

$$\left\|D_{*a}^{\alpha}G\right\|_{\infty, \prod_{i=1}^{k}[a_i^*, b_i^*]} = \left(D_{*a}^{\alpha}G\right)(x_2) > 0. \tag{1.3.30}$$

Here we define

$$JG(x) = mG(x), \quad 0 < m < \frac{1}{2}, \tag{1.3.31}$$

for any $x \in \prod_{i=1}^{k}[a_i^*, b_i^*]$.

Therefore the equation

$$JG(x) = 0, \quad x \in \prod_{i=1}^{k}[a_i^*, b_i^*], \tag{1.3.32}$$

has the same solutions as the equation

$$F(x) := \frac{JG(x)}{2D_{*a}^{\alpha}G(x_2)} = 0, \quad x \in \prod_{i=1}^{k}[a_i^*, b_i^*]. \tag{1.3.33}$$

Notice that

$$D_{*a}^{\alpha}\left(\frac{G(x)}{2D_{*a}^{\alpha}G(x_2)}\right) = \frac{D_{*a}^{\alpha}G(x)}{2D_{*a}^{\alpha}G(x_2)} \le \frac{1}{2} < 1, \quad x \in \prod_{i=1}^{k}[a_i^*, b_i^*]. \tag{1.3.34}$$

We call

$$A(x) := \frac{D^{\alpha}_{*a} G(x)}{2 D^{\alpha}_{*a} G(x_2)}, \quad \forall\, x \in \prod_{i=1}^{k} \left[a_i^*, b_i^*\right]. \tag{1.3.35}$$

We notice that

$$0 < \frac{D^{\alpha}_{*a} G(x_1)}{2 D^{\alpha}_{*a} G(x_2)} \leq A(x) \leq \frac{1}{2}. \tag{1.3.36}$$

Hence, the first condition (1.3.1) is fulfilled by

$$|1 - A(x)| = 1 - A(x) \leq 1 - \frac{D^{\alpha}_{*a} G(x_1)}{2 D^{\alpha}_{*a} G(x_2)} =: \gamma_0, \quad \forall\, x \in \prod_{i=1}^{k} \left[a_i^*, b_i^*\right]. \tag{1.3.37}$$

Hence

$$\|1 - A(x)\|_{\infty} \leq \gamma_0, \tag{1.3.38}$$

where $\|\cdot\|_{\infty}$ is over $\prod_{i=1}^{k} \left[a_i^*, b_i^*\right]$.

Clearly $\gamma_0 \in (0, 1)$.

Next we assume that $\frac{G(x)}{2(D^{\alpha}_{*a} G)(x_2)}$ is a contraction, that is

$$\left| \frac{G(x)}{2\left(D^{\alpha}_{*a} G\right)(x_2)} - \frac{G(y)}{2 D^{\alpha}_{*a} G(x_2)} \right| \leq \theta \|x - y\|, \quad \text{all } x, y \in \prod_{i=1}^{k} \left[a_i^*, b_i^*\right] \tag{1.3.39}$$

with $0 < \theta < 1$.

Hence

$$\left| \frac{mG(x)}{2\left(D^{\alpha}_{*a} G\right)(x_2)} - \frac{mG(y)}{2\left(D^{\alpha}_{*a} G\right)(x_2)} \right| \leq m\theta \|x - y\| \leq \frac{\theta}{2} \|x - y\|, \tag{1.3.40}$$

all $x, y \in \prod_{i=1}^{k} \left[a_i^*, b_i^*\right]$.

Set $\lambda = \frac{\theta}{2}$, it is $0 < \lambda < \frac{1}{2}$. We have that

$$|F(x) - F(y)| \leq \lambda \|x - y\|, \tag{1.3.41}$$

all $x, y \in \prod_{i=1}^{k} \left[a_i^*, b_i^*\right]$.

Equivalently we have

$$|JG(x) - JG(y)| \leq 2\lambda \left(D_{*a}^{\alpha} G\right)(x_2) \|x - y\|, \quad \text{all } x, y \in \prod_{i=1}^{k} \left[a_i^*, b_i^*\right]. \tag{1.3.42}$$

We observe that

$$\left\| (F(y) - F(x)) \overrightarrow{i} - A(x)(y - x) \right\| \leq$$

$$|F(y) - F(x)| + |A(x)| \|y - x\| \leq$$

$$\lambda \|y - x\| + |A(x)| \|y - x\| = (\lambda + |A(x)|) \|y - x\| =: (\psi_2) \tag{1.3.43}$$

By (1.3.27), we notice that

$$\left|D_{*a}^{\alpha} G(x)\right| \leq \frac{1}{\prod_{i=1}^{k} \Gamma(m_i - \alpha_i)}.$$

$$\left(\int_{a_1}^{x_1} \dots \int_{a_k}^{x_k} \prod_{i=1}^{k} (x_i - t_i)^{m_i - \alpha_i - 1} dt_1 \dots dt_k \right) \left\| \frac{\partial^{\sum_{i=1}^{k} m_i} G}{\partial x_1^{m_1} \dots \partial x_k^{m_k}} \right\|_{\infty}$$

$$= \frac{1}{\prod_{i=1}^{k} \Gamma(m_i - \alpha_i)} \left(\prod_{i=1}^{k} \frac{(x_i - a_i)^{m_i - \alpha_i}}{m_i - \alpha_i} \right) \left\| \frac{\partial^{\sum_{i=1}^{k} m_i} G}{\partial x_1^{m_1} \dots \partial x_k^{m_k}} \right\|_{\infty}$$

$$= \left(\prod_{i=1}^{k} \frac{(x_i - a_i)^{m_i - \alpha_i}}{\Gamma(m_i - \alpha_i + 1)} \right) \left\| \frac{\partial^{\sum_{i=1}^{k} m_i} G}{\partial x_1^{m_1} \dots \partial x_k^{m_k}} \right\|_{\infty}. \tag{1.3.44}$$

We have proved that

$$\left|D_{*a}^{\alpha} G(x)\right| \leq \left(\prod_{i=1}^{k} \frac{(b_i - a_i)^{m_i - \alpha_i}}{\Gamma(m_i - \alpha_i + 1)} \right) \left\| \frac{\partial^{\sum_{i=1}^{k} m_i} G}{\partial x_1^{m_1} \dots \partial x_k^{m_k}} \right\|_{\infty}, \tag{1.3.45}$$

$\forall\, x \in \prod_{i=1}^{k} \left[a_i^*, b_i^*\right]$, where $\|\cdot\|_{\infty}$ now is over $\prod_{i=1}^{k} [a_i, b_i]$.

Hence we get

$$|A(x)| \leq \frac{1}{2D_{*a}^{\alpha} G(x_2)} \left(\prod_{i=1}^{k} \frac{(b_i - a_i)^{m_i - \alpha_i}}{\Gamma(m_i - \alpha_i + 1)} \right) \left\| \frac{\partial^{\sum_{i=1}^{k} m_i} G}{\partial x_1^{m_1} ... \partial x_k^{m_k}} \right\|_{\infty} < \infty, \quad (1.3.46)$$

$\forall\, x \in \prod_{i=1}^{k} [a_i^*, b_i^*]$.

Therefore we obtain

$$(\psi_2) \leq \left(\lambda + \frac{1}{2D_{*a}^{\alpha} G(x_2)} \left(\prod_{i=1}^{k} \frac{(b_i - a_i)^{m_i - \alpha_i}}{\Gamma(m_i - \alpha_i + 1)} \right) \left\| \frac{\partial^{\sum_{i=1}^{k} m_i} G}{\partial x_1^{m_1} ... \partial x_k^{m_k}} \right\|_{\infty} \right) \|y - x\|, \quad (1.3.47)$$

$\forall\, x, y \in \prod_{i=1}^{k} [a_i^*, b_i^*]$.

Call

$$0 < \gamma_1 := \lambda + \frac{1}{2D_{*a}^{\alpha} G(x_2)} \left(\prod_{i=1}^{k} \frac{(b_i - a_i)^{m_i - \alpha_i}}{\Gamma(m_i - \alpha_i + 1)} \right) \left\| \frac{\partial^{\sum_{i=1}^{k} m_i} G}{\partial x_1^{m_1} ... \partial x_k^{m_k}} \right\|_{\infty}, \quad (1.3.48)$$

and by choosing $(b_i - a_i)$ small enough, $i = 1, ..., k$, we can make $\gamma_1 \in (0, 1)$, fulfilling (1.3.2).

Next we call and we need that

$$0 < \gamma := \gamma_0 + \gamma_1 = \left(1 - \frac{D_{*a}^{\alpha} G(x_1)}{2D_{*a}^{\alpha} G(x_2)} \right) +$$

$$\left\{ \lambda + \frac{1}{2D_{*a}^{\alpha} G(x_2)} \left(\prod_{i=1}^{k} \frac{(b_i - a_i)^{m_i - \alpha_i}}{\Gamma(m_i - \alpha_i + 1)} \right) \left\| \frac{\partial^{\sum_{i=1}^{k} m_i} G}{\partial x_1^{m_1} ... \partial x_k^{m_k}} \right\|_{\infty} \right\} < 1, \quad (1.3.49)$$

equivalently,

$$\lambda + \frac{1}{2D_{*a}^{\alpha} G(x_2)} \left(\prod_{i=1}^{k} \frac{(b_i - a_i)^{m_i - \alpha_i}}{\Gamma(m_i - \alpha_i + 1)} \right) \left\| \frac{\partial^{\sum_{i=1}^{k} m_i} G}{\partial x_1^{m_1} ... \partial x_k^{m_k}} \right\|_{\infty} < \frac{D_{*a}^{\alpha} G(x_1)}{2D_{*a}^{\alpha} G(x_2)}, \quad (1.3.50)$$

equivalently,

$$2\lambda D_{*a}^{\alpha}G(x_2)+\left(\prod_{i=1}^{k}\frac{(b_i-a_i)^{m_i-\alpha_i}}{\Gamma(m_i-\alpha_i+1)}\right)\left\|\frac{\partial^{\sum_{i=1}^{k}m_i}G}{\partial x_1^{m_1}...\partial x_k^{m_k}}\right\|_{\infty}<D_{*a}^{\alpha}G(x_1), \quad (1.3.51)$$

which is possible for small λ and small $(b_i - a_i)$, all $i = 1, ..., k$. That is $\gamma \in (0, 1)$, fulfilling (1.3.3). So our numerical method converges and solves (1.3.32).

References

1. S. Amat, S. Busquier, S. Plaza, Chaotic dynamics of a third-order Newton-type method. J. Math. Anal. Appl. **366**(1), 164–174 (2010)
2. G. Anastassiou, *Fractional Differentiation Inequalities* (Springer, New York, 2009)
3. G. Anastassiou, Fractional representation formulae and right fractional inequalities. Math. Comput. Model. **54**(10–12), 3098–3115 (2011)
4. G. Anastassiou, *Intelligent Mathematics: Computational Analysis* (Springer, Heidelberg, 2011)
5. G. Anastassiou, *Advanced Inequalities* (World Scientific Publisher Corp, Singapore, 2011)
6. G. Anastassiou, On left multidimensional Riemann–Liouville fractional integral. J. Comput. Anal. Appl. (2015) accepted
7. G. Anastassiou, I.K. Argyros, *Studies in Computational Intelligence, 624, Intelligent Numerical Methods: Applications to Fractional Calculus* (Springer, Heidelberg, 2016)
8. G. Anastassiou, I. Argyros, *A fixed point convergence theorem with applications in left multivariate fractional calculus* (submitted for publication, 2015)
9. I.K. Argyros, A unifying local-semilocal convergence analysis and applications for two-point Newton-like methods in Banach space. J. Math. Anal. Appl. **298**, 374–397 (2004)
10. I.K. Argyros, *Convergence and Applications of Newton-type iterations* (Springer, New York, 2008)
11. I.K. Argyros, On a class of Newton-like methods for solving nonlinear equations. J. Comput. Appl. Math. **228**, 115–122 (2009)
12. I.K. Argyros, A semilocal convergence analysis for directional Newton methods. Math. Comp., AMS **80**, 327–343 (2011)
13. I.K. Argyros, Y.J. Cho, S. Hilout, *Numerical Methods for Equations and its Applications* (CRC Press/Taylor and Fracncis Group, New York, 2012)
14. I.K. Argyros, S. Hilout, Weaker conditions for the convergence of Newton's method. J. Complex. **28**, 364–387 (2012)
15. J.A. Ezquérro, J.M. Gutiérrez, M.A. Hernández, N. Romero, M.J. Rubio, The Newton method: from Newton to Kantorovich (Spanish). Gac. R. Soc. Mat. Esp. **13**, 53–76 (2010)
16. J.A. Ezquérro, M.A. Hernández, Newton-type methods of high order and domains of semilocal and global convergence. Appl. Math. Comput. **214**(1), 142–154 (2009)
17. L.V. Kantorovich, G.P. Akilov, *Functional Analysis in Normed Spaces* (Pergamon Press, New York, 1964)

18. A.A. Magréñán, Different anomalies in a Jarratt family of iterative root finding methods. Appl. Math. Comput. **233**, 29–38 (2014)
19. A.A. Magréñán, A new tool to study real dynamics: the convergence plane. Appl. Math. Comput. **248**, 215–224 (2014)
20. F.A. Potra, V. Ptak, *Nondiscrete Induction and Iterative Processes* (Pitman Publisher, London, 1984)
21. P.D. Proinov, New general convergence theory for iterative processes and its applications to Newton–Kantorovich type theorems. J. Complex. **26**, 3–42 (2010)

Chapter 2
Fixed Point Results and Their Applications in Right Multivariate Fractional Calculus

We introduce a fixed point iterative scheme and use it to approximate a solution of a nonlinear operator equation. Applications are suggested involving in particular right multivariate fractional calculus. It follows [8].

2.1 Introduction

Let B_1, B_2 denote Banach spaces and Ω be a subset of B_1. Let also $\mathcal{L}(B_1, B_2)$ stand for the space of bounded linear operators from B_1 into B_2.

Problems in applied sciences, engineering and other disciplines can be written like

$$S(x) = 0, \tag{2.1.1}$$

where $S : \Omega \to B_2$ is a continuous operator in many cases using Mathematical Modelling [1, 7, 12, 13, 17, 19].

Solving such equations is a challenge. Closed form solutions x^* can be obtained only in some special cases. Therefore, researchers resort mostly to the utilization of iterative methods [1, 7, 12].

In the present chapter we shall approximate x^* with a very general iterative process allowing applications in diverse areas including right multivariate fractional calculus as follows: Let $A(\cdot) : \Omega \to \mathcal{L}(B_1, B_2)$ be a continuous operator and set

$$F = LS, \tag{2.1.2}$$

for some $L \in \mathcal{L}(B_2, B_1)$. The solution x^* is approximated as a limit of the sequence $\{x_n\}$ given for $x_0 \in \Omega$ by the fixed point scheme:

$$\begin{aligned} x_{n+1} &:= x_n + w_n, \ A(x_n) w_n + F(x_n) = 0 \\ \Leftrightarrow w_n &= Q(w_n) := (I - A(x_n)) w_n - F(x_n). \end{aligned} \tag{2.1.3}$$

G.A. Anastassiou and I.K. Argyros, *Intelligent Numerical Methods II: Applications to Multivariate Fractional Calculus*, Studies in Computational Intelligence 649, DOI 10.1007/978-3-319-33606-0_2

Clearly, the sequence $\{x_n\}$ given by

$$x_{n+1} = Q(x_n) = Q^{(n+1)}(x_0) \tag{2.1.4}$$

is well defined. Suppose that sequence $\{x_n\}$ converges. Then, we can write:

$$Q^{\infty}(x_0) := \lim_{n\to\infty} \left(Q^n(x_0)\right) = \lim_{n\to\infty} x_n. \tag{2.1.5}$$

Many methods in the literature can be considered special cases of method (2.1.3). We can choose A to be: $A(x) = F'(x)$ (Newton's method), $A(x) = F'(x_0)$ (Modified Newton's method), $A(x) = [x, g(x); F]$, $g : \Omega \to B_1$ (Steffensen's method). Many other choices for A can be found in [1–21] and the references there in. Therefore, it is important to study the convergence of method (2.1.3) under generalized conditions. In particular, we present the semi-local convergence of method (2.1.3) using only continuity assumptions on operator F and for a so general operator A as to allow applications to right multivariate fractional calculus and other areas.

The rest of the chapter is organized as follows: Sect. 2.2 contains the semi-local convergence of method (2.1.3). In the concluding Sect. 2.3, we suggest some applications to right multivariate fractional calculus.

2.2 Convergence

Let $B(x, \xi)$, $\overline{B}(x, \xi)$ stand, respectively for the open and closed balls in B_1 with center $x \in B_1$ and of radius $\xi > 0$.

We present the semi-local convergence of method (2.1.3) in this section.

Theorem 2.1 *Let $F : \Omega \subset B_1 \to B_2$, $A(\cdot) : \Omega \to \mathcal{L}(B_1, B_1)$ and $x_0 \in \Omega$ be as defined in the Introduction. Suppose: there exist $\delta_0 \in (0, 1)$, $\delta_1 \in (0, 1)$, $\eta \geq 0$ such that for each $x, y \in \Omega$*

$$\delta := \delta_0 + \delta_1 < 1, \tag{2.2.1}$$

$$\|F(x_0)\| \leq \eta, \tag{2.2.2}$$

$$\|I - A(x)\| \leq \delta_0, \tag{2.2.3}$$

$$\|F(y) - F(x) - A(x)(y - x)\| \leq \delta_1 \|y - x\| \tag{2.2.4}$$

and

$$\overline{B}(x_0, \delta) \subseteq \Omega, \tag{2.2.5}$$

where

$$\rho = \frac{\eta}{1-\delta}. \tag{2.2.6}$$

Then, sequence $\{x_n\}$ *generated for* $x_0 \in \Omega$ *by*

$$x_{n+1} = x_n + Q_n^{\infty}(0), \quad Q_n(w) := (I - A(x_n))\, w - F(x_n) \tag{2.2.7}$$

is well defined in $\overline{B}(x_0, \rho)$, *remains in* $\overline{B}(x_0, \rho)$ *for each* $n = 0, 1, 2, \ldots$ *and converges to* x^* *which is the only solution of equation* $F(x) = 0$ *in* $\overline{B}(x_0, \rho)$. *Moreover, an apriori error estimate is given by the sequence* $\{\rho_n\}$ *defined by*

$$\rho_0 := \rho, \quad \rho_n = T_n^{\infty}(0), \quad T_n(t) = \delta_0 + \delta_1 \rho_{n-1} \tag{2.2.8}$$

for each $n = 1, 2, \ldots$ *and satisfying*

$$\lim_{n \to \infty} \rho_n = 0. \tag{2.2.9}$$

Furthermore, an aposteriori error estimate is given by the sequence $\{\sigma_n\}$ *defined by*

$$\sigma_n := H_n^{\infty}(0), \quad H_n(t) = \delta t + \delta_1 p_{n-1}, \tag{2.2.10}$$

$$q_n := \|x_n - x_0\| \le \rho - \rho_n \le \rho, \tag{2.2.11}$$

where

$$p_{n-1} := \|x_n - x_{n-1}\| \text{ for each } n = 1, 2, \ldots \tag{2.2.12}$$

Proof We shall show using mathematical induction the following assertion is true:

(A_n) $x_n \in X$ and $\rho_n \ge 0$ are well defined and such that

$$\rho_n + p_{n-1} \le \rho_{n-1}. \tag{2.2.13}$$

By the definition of ρ, (2.2.3)–(2.2.6) we have that there exists $r \le \rho$ (Lemma 1.4 [7, pp.3]) such that

$$\delta_0 \tau + \|F(x_0)\| = r$$

and

$$\delta_0^k r \le \delta_0^k \rho \to 0 \text{ as } k \to \infty.$$

That is (Lemma 1.5 [7, pp.4]) x_1 is well defined and $p_0 \le r$.

We need the estimate:

$$T_1(\rho - r) = \delta_0(\rho - r) + \delta_1 \rho_0 =$$

$$\delta_0 \rho - \delta_0 r + \delta_1 \rho = G_0(\rho) - r = \rho - r.$$

That is (Lemma 1.4 [7, pp.3]) ρ_1 exists and satisfies

$$\rho_1 + p_0 \leq \rho - r + r = \rho = \rho_0.$$

Hence (I_0) is true. Suppose that for each $k = 1, 2, \ldots, n$, assertion (I_k) is true. We must show: x_{k+1} exists and find a bound r for p_k. Indeed, we have in turn that

$$\begin{aligned} \delta_0 \rho_k + \delta_1 (\rho_{k-1} - \rho_k) &= \delta_0 \rho_k + \delta_1 \rho_{k-1} - \delta_1 \rho_k \\ &= T_k(\rho_k) - \delta_1 \rho_k \leq \rho_k. \end{aligned}$$

That is there exists $r \leq \rho_k$ such that

$$r = \delta_0 r + \delta_1 (\rho_{k-1} - \rho_k) \text{ and } (\delta_0 + \delta_1)^i r \to 0 \tag{2.2.14}$$

as $i \to \infty$.

The induction hypothesis gives that

$$q_k \leq \sum_{m=0}^{k-1} p_m \leq \sum_{m=0}^{k-1} (\rho_m - \rho_{m+1}) = \rho - \rho_k \leq \rho,$$

so $x_k \in \overline{B}(x_0, \rho) \subseteq \Omega$ and x_1 satisfies $\|I - A(x_1)\| \leq \delta_0$ (by (2.2.3)).

Using the induction hypothesis, (2.1.3) and (2.2.4), we get

$$\begin{aligned} \|F(x_k)\| &= \|F(x_k) - F(x_{k-1}) - A(x_{k-1})(x_k - x_{k-1})\| \\ &\leq \delta_1 p_{k-1} \leq \delta_1 (\rho_{k-1} - \rho_k) \end{aligned} \tag{2.2.15}$$

leading together with (2.2.14) to:

$$\delta_0 r + \|F(x_k)\| \leq r,$$

which implies x_{k+1} exists and $p_k \leq r \leq \rho_k$. It follows from the definition of ρ_{k+1} that

$$T_{k+1}(\rho_k - r) = T_k(\rho_k) - r = \rho_k - r,$$

so ρ_{k+1} exists and satisfies

$$\rho_{k+1} + p_k \leq \rho_k - r + r = \rho_k$$

so the induction for (I_n) is completed.

Let $j \geq k$. Then, we obtain in turn that

$$\left\| x_{j+k} - x_k \right\| \leq \sum_{i=k}^{j} p_i \leq \sum_{i=k}^{j} \left(\rho_j - \rho_{j+1} \right) = \rho_k - \rho_{j+k} \leq \rho_k. \tag{2.2.16}$$

We also get using induction that

$$\rho_{k+1} = T_{k+1}\left(\rho_{k+1}\right) \leq T_{k+1}\left(\rho_k\right) \leq \delta \rho_k \leq ... \leq \delta^{k+1} \rho. \tag{2.2.17}$$

Hence, by (2.2.1) and (2.2.17) $\lim_{k \to \infty} \rho_k = 0$, so $\{x_k\}$ is a complete sequence in a Banach space X and as such it converges to some x^*. By letting $j \to \infty$ in (2.2.16), we conclude that $x^* \in \overline{B}\left(x_k, \rho_k\right)$. Moreover, by letting $k \to \infty$ in (2.2.15) and using the continuity of F we get that $F\left(x^*\right) = 0$. Notice that

$$H_k\left(\rho_k\right) \leq T_k\left(\rho_k\right) \leq \rho_k,$$

so the apriori bound exists. That is σ_k is smaller in general than ρ_k. Clearly, the conditions of the theorem are satisfied for x_k replacing x_0 (by (2.2.16)). Hence, by (2.2.8) $x^* \in \overline{B}\left(x_n, \sigma_n\right)$, which completes the proof for the aposteriori bound. ■

Remark 2.2 (a) It follows from the proof of Theorem 2.1 that the conclusions hold, if $A(\cdot)$ is replaced by a more general continuous operator $A : \Omega \to B_1$.

(b) In the next section some applications are suggested for special choices of the "A" operators with $\gamma_0 := \delta_0$ and $\gamma_1 := \delta_1$.

2.3 Applications to Right Multivariate Fractional Calculus

Our presented earlier semi-local convergence results, see Theorem 2.1, apply in the next two multivariate fractional settings given that the following inequalities are fulfilled:

$$\left\| 1 - A(x) \right\|_{\infty} \leq \gamma_0 \in (0, 1), \tag{2.3.1}$$

and

$$\left\| \left(F(y) - F(x)\right) \overrightarrow{i} - A(x)(y - x) \right\| \leq \gamma_1 \left\| y - x \right\|, \tag{2.3.2}$$

where $\gamma_0, \gamma_1 \in (0, 1)$, furthermore

$$\gamma = \gamma_0 + \gamma_1 \in (0, 1), \tag{2.3.3}$$

for all $x, y \in \prod_{i=1}^{k} \left[a_i^*, b_i^*\right]$, where $a_i < a_i^* < b_i^* < b_i$, $i = 1, \ldots, k$.

Above $\overrightarrow{i}$ is the unit vector in $\mathbb{R}^k$, $k \in \mathbb{N}$, $\left\| \overrightarrow{i} \right\| = 1$, and $\|\cdot\|$ is a norm in $\mathbb{R}^k$.

The specific functions $A(x)$, $F(x)$ will be described next.

(I) Consider the right multidimensional Riemann–Liouville fractional integral of order $\alpha = (\alpha_1, \ldots, \alpha_k)$ ($\alpha_i > 0$, $i = 1, \ldots, k$):

$$\left(I_{b-}^{\alpha} f\right)(x) = \frac{1}{\prod_{i=1}^{k} \Gamma(\alpha_i)} \int_{x_1}^{b_1} \ldots \int_{x_k}^{b_k} \prod_{i=1}^{k} (t_i - x_i)^{\alpha_i - 1} f(t_1, \ldots, t_k)\, dt_1 \ldots dt_k, \tag{2.3.4}$$

where Γ is the gamma function, $f \in L_{\infty}\left(\prod_{i=1}^{k} [a_i, b_i]\right)$, $b = (b_1, \ldots, b_k)$, and $x = (x_1, \ldots, x_k) \in \prod_{i=1}^{k} [a_i, b_i]$.

By [6], we get that $\left(I_{b-}^{\alpha} f\right)$ is a continuous function on $\prod_{i=1}^{k} [a_i, b_i]$. Furthermore by [6] we get that I_{b-}^{α} is a bounded linear operator, which is a positive operator, plus that $\left(I_{b-}^{\alpha} f\right)(b) = 0$.

In particular, $\left(I_{b-}^{\alpha} f\right)$ is continuous on $\prod_{i=1}^{k} \left[a_i^*, b_i^*\right]$.

Thus there exist $x_1, x_2 \in \prod_{i=1}^{k} \left[a_i^*, b_i^*\right]$ such that

$$\begin{aligned} \left(I_{b-}^{\alpha} f\right)(x_1) &= \min \left(I_{b-}^{\alpha} f\right)(x), \\ \left(I_{b-}^{\alpha} f\right)(x_2) &= \max \left(I_{b-}^{\alpha} f\right)(x), \end{aligned} \tag{2.3.5}$$

over all $x \in \prod_{i=1}^{k} \left[a_i^*, b_i^*\right]$.

We assume that

$$\left(I_{b-}^{\alpha} f\right)(x_1) > 0. \tag{2.3.6}$$

Hence

$$\left\| I_{b-}^{\alpha} f \right\|_{\infty, \prod_{i=1}^{k} \left[a_i^*, b_i^*\right]} = \left(I_{b-}^{\alpha} f\right)(x_2) > 0. \tag{2.3.7}$$

Here we define

$$Jf(x) = mf(x), \quad 0 < m < \frac{1}{2}, \tag{2.3.8}$$

for any $x \in \prod_{i=1}^{k} \left[a_i^*, b_i^*\right]$.

Therefore the equation

$$Jf(x) = 0, \quad x \in \prod_{i=1}^{k} \left[a_i^*, b_i^*\right], \tag{2.3.9}$$

has the same solutions as the equation

$$F(x) := \frac{Jf(x)}{2\left(I_{b-}^{\alpha} f\right)(x_2)} = 0, \quad x \in \prod_{i=1}^{k} \left[a_i^*, b_i^*\right]. \tag{2.3.10}$$

Notice that

$$I_{b-}^{\alpha}\left(\frac{f}{2\left(I_{b-}^{\alpha} f\right)(x_2)}\right)(x) = \frac{\left(I_{b-}^{\alpha} f\right)(x)}{2\left(I_{b-}^{\alpha} f\right)(x_2)} \leq \frac{1}{2} < 1, \ x \in \prod_{i=1}^{k} \left[a_i^*, b_i^*\right]. \tag{2.3.11}$$

Call

$$A(x) := \frac{\left(I_{b-}^{\alpha} f\right)(x)}{2\left(I_{b-}^{\alpha} f\right)(x_2)}, \quad \forall x \in \prod_{i=1}^{k} \left[a_i^*, b_i^*\right]. \tag{2.3.12}$$

We notice that

$$0 < \frac{\left(I_{b-}^{\alpha} f\right)(x_1)}{2\left(I_{b-}^{\alpha} f\right)(x_2)} \leq A(x) \leq \frac{1}{2}, \quad \forall x \in \prod_{i=1}^{k} \left[a_i^*, b_i^*\right]. \tag{2.3.13}$$

Hence the first condition (2.3.1) is fulfilled by

$$|1 - A(x)| = 1 - A(x) \leq 1 - \frac{\left(I_{b-}^{\alpha} f\right)(x_1)}{2\left(I_{b-}^{\alpha} f\right)(x_2)} =: \gamma_0, \quad \forall x \in \prod_{i=1}^{k} \left[a_i^*, b_i^*\right]. \tag{2.3.14}$$

Hence $\|1 - A(x)\|_{\infty} \leq \gamma_0$, where $\|\cdot\|_{\infty}$ is over $\prod_{i=1}^{k} \left[a_i^*, b_i^*\right]$. Clearly $\gamma_0 \in (0, 1)$.

Next we assume that $\frac{f(x)}{2\left(I_{b-}^{\alpha} f\right)(x_2)}$ is a contraction, that is

$$\left|\frac{f(x)}{2\left(I_{b-}^{\alpha} f\right)(x_2)} - \frac{f(y)}{2\left(I_{b-}^{\alpha} f\right)(x_2)}\right| \leq \theta \|x - y\|, \quad \text{all } x, y \in \prod_{i=1}^{k} \left[a_i^*, b_i^*\right], \tag{2.3.15}$$

$0 < \theta < 1$.

Hence

$$\left|\frac{mf(x)}{2\left(I_{b-}^{\alpha} f\right)(x_2)} - \frac{mf(y)}{2\left(I_{b-}^{\alpha} f\right)(x_2)}\right| \leq m\theta \|x - y\| \leq \frac{\theta}{2} \|x - y\|, \tag{2.3.16}$$

all $x, y \in \prod_{i=1}^{k} \left[a_i^*, b_i^*\right]$.

Set $\lambda = \frac{\theta}{2}$, it is $0 < \lambda < \frac{1}{2}$. We have that

$$|F(x) - F(y)| \leq \lambda \|x - y\|, \tag{2.3.17}$$

all $x, y \in \prod_{i=1}^{k} [a_i^*, b_i^*]$.

Equivalently we have

$$|Jf(x) - Jf(y)| \leq 2\lambda \left(I_{b-}^{\alpha} f\right)(x_2) \|x - y\|, \text{ all } x, y \in \prod_{i=1}^{k} [a_i^*, b_i^*]. \tag{2.3.18}$$

We observe that

$$\left\|(F(y) - F(x)) \vec{i} - A(x)(y - x)\right\| \leq$$
$$|F(y) - F(x)| + |A(x)| \|y - x\| \leq \tag{2.3.19}$$
$$\lambda \|y - x\| + |A(x)| \|y - x\| = (\lambda + |A(x)|) \|y - x\| =: (\psi_1),$$

$\forall\, x, y \in \prod_{i=1}^{k} [a_i^*, b_i^*]$.

By [6], we have that

$$\left|\left(I_{b-}^{\alpha} f\right)(x)\right| \leq \left(\prod_{i=1}^{k} \frac{(b_i - a_i)^{\alpha_i}}{\Gamma(\alpha_i + 1)}\right) \|f\|_{\infty}, \tag{2.3.20}$$

$\forall\, x \in \prod_{i=1}^{k} [a_i^*, b_i^*]$, where $\|\cdot\|_{\infty}$ now is over $\prod_{i=1}^{k} [a_i, b_i]$.

Hence

$$|A(x)| = \frac{\left|\left(I_{b-}^{\alpha} f\right)(x)\right|}{2\left(I_{b-}^{\alpha} f\right)(x_2)} \leq \frac{1}{2\left(I_{b-}^{\alpha} f\right)(x_2)} \left(\prod_{i=1}^{k} \frac{(b_i - a_i)^{\alpha_i}}{\Gamma(\alpha_i + 1)}\right) \|f\|_{\infty} < \infty, \tag{2.3.21}$$

$\forall\, x \in \prod_{i=1}^{k} [a_i^*, b_i^*]$.

Therefore we get

$$(\psi_1) \leq \left(\lambda + \frac{1}{2\left(I_{b-}^{\alpha} f\right)(x_2)} \left(\prod_{i=1}^{k} \frac{(b_i - a_i)^{\alpha_i}}{\Gamma(\alpha_i + 1)}\right) \|f\|_{\infty}\right) \|y - x\|, \tag{2.3.22}$$

$\forall\, x, y \in \prod_{i=1}^{k} [a_i^*, b_i^*]$.

Call

$$0 < \gamma_1 := \lambda + \frac{1}{2\left(I_{b-}^{\alpha} f\right)(x_2)} \left(\prod_{i=1}^{k} \frac{(b_i - a_i)^{\alpha_i}}{\Gamma(\alpha_i + 1)} \right) \|f\|_{\infty}, \tag{2.3.23}$$

and by choosing $(b_i - a_i)$ small enough, $i = 1, \ldots, k$, we can make $\gamma_1 \in (0, 1)$, fulfilling (2.3.2).

Next we call and we need that

$$0 < \gamma := \gamma_0 + \gamma_1 = \left(1 - \frac{\left(I_{b-}^{\alpha} f\right)(x_1)}{2\left(I_{b-}^{\alpha} f\right)(x_2)} \right) +$$

$$\left(\lambda + \frac{1}{2\left(I_{b-}^{\alpha} f\right)(x_2)} \left(\prod_{i=1}^{k} \frac{(b_i - a_i)^{\alpha_i}}{\Gamma(\alpha_i + 1)} \right) \|f\|_{\infty} \right) < 1, \tag{2.3.24}$$

equivalently,

$$\lambda + \frac{1}{2\left(I_{b-}^{\alpha} f\right)(x_2)} \left(\prod_{i=1}^{k} \frac{(b_i - a_i)^{\alpha_i}}{\Gamma(\alpha_i + 1)} \right) \|f\|_{\infty} < \frac{\left(I_{b-}^{\alpha} f\right)(x_1)}{2\left(I_{b-}^{\alpha} f\right)(x_2)}, \tag{2.3.25}$$

equivalently,

$$2\lambda \left(I_{b-}^{\alpha} f\right)(x_2) + \left(\prod_{i=1}^{k} \frac{(b_i - a_i)^{\alpha_i}}{\Gamma(\alpha_i + 1)} \right) \|f\|_{\infty} < \left(I_{b-}^{\alpha} f\right)(x_1), \tag{2.3.26}$$

which is possible for small λ and small $(b_i - a_i)$, all $i = 1, \ldots, k$. That is $\gamma \in (0, 1)$, fulfilling (2.3.3). So our numerical method converges and solves (2.3.9).

(II) Let $\alpha = (\alpha_1, \ldots, \alpha_k)$, $\alpha_i > 0$, $m_i = \lceil \alpha_i \rceil$ ($\lceil \cdot \rceil$ ceiling function), $\alpha_i \notin \mathbb{N}$, $i = 1, \ldots, k \in \mathbb{N}$, and $G \in C^{\sum_{i=1}^{k} m_i - 1}\left(\prod_{i=1}^{k} [a_i, b_i] \right)$, such that

$$0 \neq \frac{\partial^{\sum_{i=1}^{k} m_i} G}{\partial x_1^{m_1} \ldots \partial x_k^{m_k}} \in L_{\infty}\left(\prod_{i=1}^{k} [a_i, b_i] \right).$$

Here we consider the multivariate right Caputo type fractional mixed partial derivative of order α:

$$D_{b-}^{\alpha} G(x) = \frac{(-1)^{\sum_{i=1}^{k} m_i}}{\prod_{i=1}^{k} \Gamma(m_i - \alpha_i)} \int_{x_1}^{b_1} \ldots \int_{x_k}^{b_k} \prod_{i=1}^{k} (t_i - x_i)^{m_i - \alpha_i - 1} \cdot \tag{2.3.27}$$

$$\frac{\partial^{\sum_{i=1}^{k} m_i} G(t_1, \dots, t_k)}{\partial t_1^{m_1} \dots \partial t_k^{m_k}} dt_1 \dots dt_k,$$

where again Γ is the gamma function, $b = (b_1, \dots, b_k)$, $\forall\, x = (x_1, \dots, x_k) \in \prod_{i=1}^{k} [a_i, b_i]$. Notice here that $m_i - \alpha_i > 0$, $i = 1, \dots, k$.

By [6], we get that $D_{b-}^{\alpha} G$ is a continuous function on $\prod_{i=1}^{k} [a_i, b_i]$, and it holds that $D_{b-}^{\alpha} G(b) = 0$.

In particular $D_{b-}^{\alpha} G$ is continuous on $\prod_{i=1}^{k} [a_i^*, b_i^*]$, where $a_i < a_i^* < b_i^* < b_i$, $i = 1, \dots, k$.

Therefore there exist $x_1, x_2 \in \prod_{i=1}^{k} [a_i^*, b_i^*]$ such that

$$\begin{aligned} \left(D_{b-}^{\alpha} G\right)(x_1) &= \min \left(D_{b-}^{\alpha} G\right)(x), \\ \left(D_{b-}^{\alpha} G\right)(x_2) &= \max \left(D_{b-}^{\alpha} G\right)(x), \end{aligned} \tag{2.3.28}$$

over all $x \in \prod_{i=1}^{k} [a_i^*, b_i^*]$.

We assume that

$$\left(D_{b-}^{\alpha} G\right)(x_1) > 0. \tag{2.3.29}$$

Hence

$$\left\| D_{b-}^{\alpha} G \right\|_{\infty, \prod_{i=1}^{k} [a_i^*, b_i^*]} = \left(D_{b-}^{\alpha} G\right)(x_2) > 0. \tag{2.3.30}$$

Here we define

$$JG(x) = mG(x), \quad 0 < m < \frac{1}{2}, \tag{2.3.31}$$

for any $x \in \prod_{i=1}^{k} [a_i^*, b_i^*]$.

Therefore the equation

$$JG(x) = 0, \quad x \in \prod_{i=1}^{k} [a_i^*, b_i^*], \tag{2.3.32}$$

has the same solutions as the equation

$$F(x) := \frac{JG(x)}{2 D_{b-}^{\alpha} G(x_2)} = 0, \quad x \in \prod_{i=1}^{k} [a_i^*, b_i^*]. \tag{2.3.33}$$

Notice that

$$D_{b-}^{\alpha}\left(\frac{G(x)}{2D_{b-}^{\alpha}G(x_2)}\right)=\frac{D_{b-}^{\alpha}G(x)}{2D_{b-}^{\alpha}G(x_2)}\leq\frac{1}{2}<1,\quad x\in\prod_{i=1}^{k}\left[a_i^*,b_i^*\right]. \tag{2.3.34}$$

We call

$$A(x):=\frac{D_{b-}^{\alpha}G(x)}{2D_{b-}^{\alpha}G(x_2)},\quad \forall x\in\prod_{i=1}^{k}\left[a_i^*,b_i^*\right]. \tag{2.3.35}$$

We notice that

$$0<\frac{D_{b-}^{\alpha}G(x_1)}{2D_{b-}^{\alpha}G(x_2)}\leq A(x)\leq\frac{1}{2}. \tag{2.3.36}$$

Hence the first condition (2.3.1) is fulfilled by

$$|1-A(x)|=1-A(x)\leq 1-\frac{D_{b-}^{\alpha}G(x_1)}{2D_{b-}^{\alpha}G(x_2)}=:\gamma_0,\quad \forall x\in\prod_{i=1}^{k}\left[a_i^*,b_i^*\right]. \tag{2.3.37}$$

Hence

$$\|1-A(x)\|_{\infty}\leq\gamma_0, \tag{2.3.38}$$

where $\|\cdot\|_{\infty}$ is over $\prod_{i=1}^{k}\left[a_i^*,b_i^*\right]$.

Clearly $\gamma_0\in(0,1)$.

Next we assume that $\frac{G(x)}{2\left(D_{b-}^{\alpha}G\right)(x_2)}$ is a contraction, that is

$$\left|\frac{G(x)}{2\left(D_{b-}^{\alpha}G\right)(x_2)}-\frac{G(y)}{2D_{b-}^{\alpha}G(x_2)}\right|\leq\theta\|x-y\|,\ \text{all}\ x,y\in\prod_{i=1}^{k}\left[a_i^*,b_i^*\right], \tag{2.3.39}$$

with $0<\theta<1$.

Hence

$$\left|\frac{mG(x)}{2\left(D_{b-}^{\alpha}G\right)(x_2)}-\frac{mG(y)}{2\left(D_{b-}^{\alpha}G\right)(x_2)}\right|\leq m\theta\|x-y\|\leq\frac{\theta}{2}\|x-y\|, \tag{2.3.40}$$

all $x,y\in\prod_{i=1}^{k}\left[a_i^*,b_i^*\right]$.

Set $\lambda=\frac{\theta}{2}$, it is $0<\lambda<\frac{1}{2}$. We have that

$$|F(x)-F(y)|\leq\lambda\|x-y\|, \tag{2.3.41}$$

all $x, y \in \prod_{i=1}^{k} [a_i^*, b_i^*]$.

Equivalently we have

$$|JG(x) - JG(y)| \leq 2\lambda \left(D_{b-}^{\alpha} G\right)(x_2) \|x - y\|, \text{ all } x, y \in \prod_{i=1}^{k} [a_i^*, b_i^*]. \quad (2.3.42)$$

We observe that

$$\left\| (F(y) - F(x)) \overrightarrow{i} - A(x)(y - x) \right\| \leq$$

$$|F(y) - F(x)| + |A(x)| \|y - x\| \leq$$

$$\lambda \|y - x\| + |A(x)| \|y - x\| = (\lambda + |A(x)|) \|y - x\| =: (\psi_2), \quad (2.3.43)$$

$\forall\, x, y \in \prod_{i=1}^{k} [a_i^*, b_i^*]$.

By (2.3.27), we notice that

$$\left|D_{b-}^{\alpha} G(x)\right| \leq \frac{1}{\prod_{i=1}^{k} \Gamma(m_i - \alpha_i)}.$$

$$\left(\int_{x_1}^{b_1} \ldots \int_{x_k}^{b_k} \prod_{i=1}^{k} (t_i - x_i)^{m_i - \alpha_i - 1} dt_1 \ldots dt_k\right) \left\| \frac{\partial^{\sum_{i=1}^{k} m_i} G}{\partial x_1^{m_1} \ldots \partial x_k^{m_k}} \right\|_{\infty}$$

$$= \frac{1}{\prod_{i=1}^{k} \Gamma(m_i - \alpha_i)} \left(\prod_{i=1}^{k} \frac{(b_i - x_i)^{m_i - \alpha_i}}{m_i - \alpha_i}\right) \left\| \frac{\partial^{\sum_{i=1}^{k} m_i} G}{\partial x_1^{m_1} \ldots \partial x_k^{m_k}} \right\|_{\infty}$$

$$= \left(\prod_{i=1}^{k} \frac{(b_i - x_i)^{m_i - \alpha_i}}{\Gamma(m_i - \alpha_i + 1)}\right) \left\| \frac{\partial^{\sum_{i=1}^{k} m_i} G}{\partial x_1^{m_1} \ldots \partial x_k^{m_k}} \right\|_{\infty}. \quad (2.3.44)$$

We have proved that

$$\left|D_{b-}^{\alpha} G(x)\right| \leq \left(\prod_{i=1}^{k} \frac{(b_i - a_i)^{m_i - \alpha_i}}{\Gamma(m_i - \alpha_i + 1)}\right) \left\| \frac{\partial^{\sum_{i=1}^{k} m_i} G}{\partial x_1^{m_1} \ldots \partial x_k^{m_k}} \right\|_{\infty}, \quad (2.3.45)$$

$\forall\, x \in \prod_{i=1}^{k} [a_i^*, b_i^*]$, where $\|\cdot\|_\infty$ now is over $\prod_{i=1}^{k} [a_i, b_i]$.

Hence we get

$$|A(x)| \leq \frac{1}{2D_{b-}^{\alpha} G(x_2)} \left(\prod_{i=1}^{k} \frac{(b_i - a_i)^{m_i - \alpha_i}}{\Gamma(m_i - \alpha_i + 1)} \right) \left\| \frac{\partial^{\sum_{i=1}^{k} m_i} G}{\partial x_1^{m_1} \ldots \partial x_k^{m_k}} \right\|_\infty < \infty, \tag{2.3.46}$$

$\forall\, x \in \prod_{i=1}^{k} [a_i^*, b_i^*]$.

Therefore we obtain

$$(\psi_2) \leq \left(\lambda + \frac{1}{2D_{b-}^{\alpha} G(x_2)} \left(\prod_{i=1}^{k} \frac{(b_i - a_i)^{m_i - \alpha_i}}{\Gamma(m_i - \alpha_i + 1)} \right) \left\| \frac{\partial^{\sum_{i=1}^{k} m_i} G}{\partial x_1^{m_1} \ldots \partial x_k^{m_k}} \right\|_\infty \right) \|y - x\|, \tag{2.3.47}$$

$\forall\, x, y \in \prod_{i=1}^{k} [a_i^*, b_i^*]$.

Call

$$0 < \gamma_1 := \lambda + \frac{1}{2D_{b-}^{\alpha} G(x_2)} \left(\prod_{i=1}^{k} \frac{(b_i - a_i)^{m_i - \alpha_i}}{\Gamma(m_i - \alpha_i + 1)} \right) \left\| \frac{\partial^{\sum_{i=1}^{k} m_i} G}{\partial x_1^{m_1} \ldots \partial x_k^{m_k}} \right\|_\infty, \tag{2.3.48}$$

and by choosing $(b_i - a_i)$ small enough, $i = 1, \ldots, k$, we can make $\gamma_1 \in (0, 1)$, fulfilling (2.3.2).

Next we call and we need that

$$0 < \gamma := \gamma_0 + \gamma_1 = \left(1 - \frac{D_{b-}^{\alpha} G(x_1)}{2D_{b-}^{\alpha} G(x_2)} \right) +$$

$$\left\{ \lambda + \frac{1}{2D_{b-}^{\alpha} G(x_2)} \left(\prod_{i=1}^{k} \frac{(b_i - a_i)^{m_i - \alpha_i}}{\Gamma(m_i - \alpha_i + 1)} \right) \left\| \frac{\partial^{\sum_{i=1}^{k} m_i} G}{\partial x_1^{m_1} \ldots \partial x_k^{m_k}} \right\|_\infty \right\} < 1, \tag{2.3.49}$$

equivalently,

$$\lambda + \frac{1}{2D_{b-}^{\alpha} G(x_2)} \left(\prod_{i=1}^{k} \frac{(b_i - a_i)^{m_i - \alpha_i}}{\Gamma(m_i - \alpha_i + 1)} \right) \left\| \frac{\partial^{\sum_{i=1}^{k} m_i} G}{\partial x_1^{m_1} \ldots \partial x_k^{m_k}} \right\|_\infty < \frac{D_{b-}^{\alpha} G(x_1)}{2D_{b-}^{\alpha} G(x_2)}, \tag{2.3.50}$$

equivalently,

$$2\lambda D_{b-}^{\alpha} G(x_2) + \left(\prod_{i=1}^{k} \frac{(b_i - a_i)^{m_i - \alpha_i}}{\Gamma(m_i - \alpha_i + 1)} \right) \left\| \frac{\partial^{\sum_{i=1}^{k} m_i} G}{\partial x_1^{m_1} \ldots \partial x_k^{m_k}} \right\|_{\infty} < D_{b-}^{\alpha} G(x_1), \tag{2.3.51}$$

which is possible for small λ and small $(b_i - a_i)$, all $i = 1, \ldots, k$. That is $\gamma \in (0, 1)$, fulfilling (2.3.3). So our numerical method converges and solves (2.3.32).

References

1. S. Amat, S. Busquier, S. Plaza, Chaotic dynamics of a third-order Newton-type method. J. Math. Anal. Appl. **366**(1), 164–174 (2010)
2. G. Anastassiou, *Fractional Differentiation Inequalities* (Springer, New York, 2009)
3. G. Anastassiou, Fractional representation formulae and right fractional inequalities. Math. Comput. Model. **54**(10–12), 3098–3115 (2011)
4. G. Anastassiou, *Intelligent Mathematics: Computational Analysis* (Springer, Heidelberg, 2011)
5. G. Anastassiou, *Advanced Inequalities* (World Scientific Publ. Corp, Singapore, 2011)
6. G. Anastassiou, On right multidimensional Riemann-Liouville fractional integral. J. Comput. Anal. Appl. (2015)
7. G. Anastassiou, I.K. Argyros, *Intelligent Numerical Methods: Applications to Fractional Calculus*, Studies in Computational Intelligence (Springer, Heidelberg, 2016)
8. G. Anastassiou, I. Argyros, Fixed point schemes with applications in right multivariate fractional calculus. submitted for publication (2015)
9. I.K. Argyros, A unifying local-semilocal convergence analysis and applications for two-point Newton-like methods in Banach space. J. Math. Anal. Appl. **298**, 374–397 (2004)
10. I.K. Argyros, *Convergence and Applications of NewtonType Iterations* (Springer, New York, 2008)
11. I.K. Argyros, On a class of Newton-like methods for solving nonlinear equations. J. Comput. Appl. Math. **228**, 115–122 (2009)
12. I.K. Argyros, A semilocal convergence analysis for directional Newton methods. AMS J. **80**, 327–343 (2011)
13. I.K. Argyros, Y.J. Cho, S. Hilout, *Numerical Methods for Equations and Its Applications* (CRC Press/Taylor and Fracncis, New York, 2012)
14. I.K. Argyros, S. Hilout, Weaker conditions for the convergence of Newton's method. J. Complex. **28**, 364–387 (2012)
15. J.A. Ezquérro, J.M. Gutiérrez, M.A. Hernández, N. Romero, M.J. Rubio, The Newton method: from Newton to Kantorovich (Spanish). Gac. R. Soc. Mat. Esp. **13**, 53–76 (2010)
16. J.A. Ezquérro, M.A. Hernández, Newton-type methods of high order and domains of semilocal and global convergence. Appl. Math. Comput. **214**(1), 142–154 (2009)
17. L.V. Kantorovich, G.P. Akilov, *Functional Analysis in Normed Spaces* (Pergamon Press, New York, 1964)
18. A.A. Magreñán, Different anomalies in a Jarratt family of iterative root finding methods. Appl. Math. Comput. **233**, 29–38 (2014)
19. A.A. Magreñán, A new tool to study real dynamics: the convergence plane. Appl. Math. Comput. **248**, 215–224 (2014)
20. F.A. Potra, V. Ptak, *Nondiscrete Induction and Iterative Processes* (Pitman, London, 1984)
21. P.D. Proinov, New general convergence theory for iterative processes and its applications to Newton-Kantorovich type theorems. J. Complex. **26**, 3–42 (2010)

Chapter 3
Semi-local Convergence of Iterative Procedures and Their Applications in k-Multivariate Fractional Calculus

We provide new semi-local convergence results for general iterative methods in order to approximate a solution of a nonlinear operator equation. Moreover, applications are suggested in many areas including k-multivariate fractional calculus, where k is a positive integer. It follows [8].

3.1 Introduction

Many problems are special cases of the equation

$$M(x) = 0, \tag{3.1.1}$$

where $M : \Omega \to B_2$ is a continuous operator, B_1, B_2 are Banach spaces and $\Omega \subseteq B_1$. These problems are reduced to (3.1.1) using Mathematical Modelling.

Then, it is very important to find solutions x^* of Eq. (3.1.1). However, the solutions x^* can rarely be obtained in closed form. That is why we use mostly iterative methods to approximate such solutions [1, 8–21].

Let $\mathcal{L}(B_1, B_2)$ stand for space of bounded linear operators from B_1 into B_2. Let also $A(\cdot) : \Omega \to \mathcal{L}(B_1, B_1)$ be a continuous operator. Set

$$F = LM, \tag{3.1.2}$$

where $L \in \mathcal{L}(B_2, B_1)$. We shall approximate x^* using a sequence $\{x_n\}$ generated by the fixed point scheme:

$$\begin{aligned} x_{n+1} &:= x_n + y_n, \ A(x_n) y_n + F(x_n) = 0 \\ \Leftrightarrow y_n &= Q(y_n) := (I - A(x_n)) y_n - F(x_n), \end{aligned} \tag{3.1.3}$$

G.A. Anastassiou and I.K. Argyros, *Intelligent Numerical Methods II: Applications to Multivariate Fractional Calculus*, Studies in Computational Intelligence 649, DOI 10.1007/978-3-319-33606-0_3

where $x_0 \in \Omega$. The sequence $\{x_n\}$ defined by

$$x_{n+1} = Q(x_n) = Q^{(n+1)}(x_0) \tag{3.1.4}$$

exists. In case of convergence we write:

$$Q^{\infty}(x_0) := \lim_{n\to\infty} \left(Q^n(x_0)\right) = \lim_{n\to\infty} x_n. \tag{3.1.5}$$

Many methods in the literature can be considered special cases of method (3.1.3). We can choose A to be: $A(x) = F'(x)$ (Newton's method), $A(x) = F'(x_0)$ (Modified Newton's method), $A(x) = [x, g(x); F]$, $g : \Omega \to B_1$ (Steffensen's method). Many other choices for A can be found in [1–22] and the references there in. Therefore, it is important to study the convergence of method (3.1.3) under generalized conditions. In particular, we present the semi-local convergence of method (3.1.3) using only continuity assumptions on operator F and for a so general operator A as to allow applications to k-multivariate fractional calculus and other areas.

The rest of the chapter is organized as follows: Sect. 3.2 contains the semi-local convergence of method (3.1.3). In the concluding Sect. 3.3, we suggest some applications to k-multivariate fractional calculus.

3.2 Convergence

Let $B(w, \xi)$, $\overline{B}(w, \xi)$ stand, respectively for the open and closed balls in B_1 with center $w \in B_1$ and of radius $\xi > 0$.

We present the semi-local convergence of method (3.1.3) in this section.

Theorem 3.1 *Let $F : \Omega \subset B_1 \to B_2$, $A(\cdot) : \Omega \to \mathcal{L}(B_1, B_1)$ and $x_0 \in \Omega$ be as defined in the Introduction. Suppose: there exist $\delta_0 \in (0, 1)$, $\delta_1 \in (0, 1)$, $\eta \geq 0$ such that for each $x, y \in \Omega$*

$$\delta := \delta_0 + \delta_1 < 1, \tag{3.2.1}$$

$$\|F(x_0)\| \leq \eta, \tag{3.2.2}$$

$$\|I - A(x)\| \leq \delta_0, \tag{3.2.3}$$

$$\|F(y) - F(x) - A(x)(y - x)\| \leq \delta_1 \|y - x\| \tag{3.2.4}$$

and

$$\overline{B}(x_0, \delta) \subseteq \Omega, \tag{3.2.5}$$

where

$$\rho = \frac{\eta}{1 - \delta}. \tag{3.2.6}$$

Then, sequence $\{x_n\}$ *generated for* $x_0 \in \Omega$ *by*

$$x_{n+1} = x_n + Q_n^{\infty}(0), \quad Q_n(y) := (I - A(x_n))\, y - F(x_n) \tag{3.2.7}$$

is well defined in $B(x_0, \rho)$*, remains in* $\overline{B}(x_0, \rho)$ *for each* $n = 0, 1, 2, \ldots$ *and converges to* x^* *which is the only solution of equation* $F(x) = 0$ *in* $\overline{B}(x_0, \rho)$*. Moreover, an apriori error estimate is given by the sequence* $\{\rho_n\}$ *defined by*

$$\rho_0 := \rho, \; \rho_n = T_n^{\infty}(0), \; T_n(t) = \delta_0 + \delta_1 \rho_{n-1} \tag{3.2.8}$$

for each $n = 1, 2, \ldots$ *and satisfying*

$$\lim_{n\to\infty} \rho_n = 0. \tag{3.2.9}$$

Furthermore, an aposteriori error estimate is given by the sequence $\{\sigma_n\}$ *defined by*

$$\sigma_n := H_n^{\infty}(0), \; H_n(t) = \delta t + \delta_1 p_{n-1}, \tag{3.2.10}$$

$$q_n := \|x_n - x_0\| \leq \rho - \rho_n \leq \rho, \tag{3.2.11}$$

where

$$p_{n-1} := \|x_n - x_{n-1}\| \text{ for each } n = 1, 2, \ldots \tag{3.2.12}$$

Proof We shall show using mathematical induction the following assertion is true:
(A_n) $x_n \in X$ and $\rho_n \geq 0$ are well defined and such that

$$\rho_n + p_{n-1} \leq \rho_{n-1}. \tag{3.2.13}$$

By the definition of ρ, (3.2.3)–(3.2.6) we have that there exists $r \leq \rho$ (Lemma 1.4 [9, p. 3]) such that

$$\delta_0 r + \|F(x_0)\| = r$$

and

$$\delta_0^k r \leq \delta_0^k \rho \to 0 \text{ as } k \to \infty.$$

That is (Lemma 1.5 [9, p. 4]) x_1 is well defined and $p_0 \leq r$.
We need the estimate:

$$T_1(\rho - r) = \delta_0(\rho - r) + \delta_1 \rho_0 =$$

$$\delta_0 \rho - \delta_0 r + \delta_1 \rho = G_0(\rho) - r = \rho - r.$$

That is (Lemma 1.4 [9, p. 3]) ρ_1 exists and satisfies

$$\rho_1 + p_0 \leq \rho - r + r = \rho = \rho_0.$$

Hence (I_0) is true. Suppose that for each $k = 1, 2, ..., n,$ assertion (I_k) is true. We must show: x_{k+1} exists and find a bound r for p_k. Indeed, we have in turn that

$$\begin{aligned} \delta_0 \rho_k + \delta_1 (\rho_{k-1} - \rho_k) &= \delta_0 \rho_k + \delta_1 \rho_{k-1} - \delta_1 \rho_k \\ &= T_k (\rho_k) - \delta_1 \rho_k \leq \rho_k. \end{aligned}$$

That is there exists $r \leq \rho_k$ such that

$$r = \delta_0 r + \delta_1 (\rho_{k-1} - \rho_k) \text{ and } (\delta_0 + \delta_1)^i r \to 0 \tag{3.2.14}$$

as $i \to \infty$.

The induction hypothesis gives that

$$q_k \leq \sum_{m=0}^{k-1} p_m \leq \sum_{m=0}^{k-1} (\rho_m - \rho_{m+1}) = \rho - \rho_k \leq \rho,$$

so $x_k \in \overline{B} (x_0, \rho) \subseteq \Omega$ and x_1 satisfies $\|I - A (x_1)\| \leq \delta_0$ (by (3.2.3)).

Using the induction hypothesis, (3.1.3) and (3.2.4), we get

$$\begin{aligned} \|F (x_k)\| &= \|F (x_k) - F (x_{k-1}) - A (x_{k-1}) (x_k - x_{k-1})\| \\ &\leq \delta_1 p_{k-1} \leq \delta_1 (\rho_{k-1} - \rho_k) \end{aligned} \tag{3.2.15}$$

leading together with (3.2.14) to:

$$\delta_0 r + \|F (x_k)\| \leq r,$$

which implies x_{k+1} exists and $p_k \leq r \leq \rho_k$. It follows from the definition of ρ_{k+1} that

$$T_{k+1} (\rho_k - r) = T_k (\rho_k) - r = \rho_k - r,$$

so ρ_{k+1} exists and satisfies

$$\rho_{k+1} + p_k \leq \rho_k - r + r = \rho_k$$

so the induction for (I_n) is completed.

Let $j \geq k$. Then, we obtain in turn that

$$\left\|x_{j+k} - x_k\right\| \leq \sum_{i=k}^{j} p_i \leq \sum_{i=k}^{j} \left(\rho_j - \rho_{j+1}\right) = \rho_k - \rho_{j+k} \leq \rho_k. \tag{3.2.16}$$

We also obtain using induction that

$$\rho_{k+1} = T_{k+1}\left(\rho_{k+1}\right) \leq T_{k+1}\left(\rho_k\right) \leq \delta \rho_k \leq \ldots \leq \delta^{k+1} \rho. \tag{3.2.17}$$

Hence, by (3.2.1) and (3.2.17) $\lim_{k \to \infty} \rho_k = 0$, so $\{x_k\}$ is a complete sequence in a Banach space X and as such it converges to some x^*. By letting $j \to \infty$ in (3.2.16), we conclude that $x^* \in \overline{B}\left(x_k, \rho_k\right)$. Moreover, by letting $k \to \infty$ in (3.2.15) and using the continuity of F we get that $F\left(x^*\right) = 0$. Notice that

$$H_k\left(\rho_k\right) \leq T_k\left(\rho_k\right) \leq \rho_k,$$

so the apriori bound exists. That is σ_k is smaller in general than ρ_k. Clearly, the conditions of the theorem are satisfied for x_k replacing x_0 (by (3.2.16)). Hence, by (3.2.8) $x^* \in \overline{B}\left(x_n, \sigma_n\right)$, which completes the proof for the aposteriori bound. ■

Remark 3.2 (a) It follows from the proof of Theorem 3.1 that the conclusions hold, if $A(\cdot)$ is replaced by a more general continuous operator $A : \Omega \to B_1$.

(b) In the next section some applications are suggested for special choices of the "A" operators with $\gamma_0 := \delta_0$ and $\gamma_1 := \delta_1$.

3.3 Applications to k-Multivariate Fractional Calculus

Our presented earlier semi-local convergence results, see Theorem 3.1, apply in the next three multivariate fractional settings given that the following inequalities are fulfilled:

$$\|1 - A(x)\|_{\infty} \leq \gamma_0 \in (0, 1), \tag{3.3.1}$$

and

$$\left\|(F(y) - F(x)) \overrightarrow{i} - A(x)(y - x)\right\| \leq \gamma_1 \|y - x\|, \tag{3.3.2}$$

where $\gamma_0, \gamma_1 \in (0, 1)$, furthermore

$$\gamma = \gamma_0 + \gamma_1 \in (0, 1), \tag{3.3.3}$$

for all $x, y \in \prod_{i=1}^{N} \left[a_i^*, b_i^*\right]$, where $a_i < a_i^* < b_i^* < b_i$, $i = 1, \ldots, N$.

Above $\overrightarrow{i}$ is the unit vector in $\mathbb{R}^N$, $N \in \mathbb{N}$, $\left\|\overrightarrow{i}\right\| = 1$, and $\|\cdot\|$ is a norm in $\mathbb{R}^k$.

The specific functions $A(x)$, $F(x)$ will be described next.

(I) Consider the k-left multidimensional Riemann–Liouville fractional integral of order $\alpha = (\alpha_1, ..., \alpha_N)$, $k = (k_1, ..., k_N)$, $(\alpha_i > 0, k_i > 0, i = 1, ..., N)$:

$$\left({}_kI_{a+}^{\alpha} f\right)(x) = \frac{1}{\prod_{i=1}^{N} k_i \Gamma_{k_i}(\alpha_i)} \int_{a_1}^{x_1} \cdots \int_{a_N}^{x_N} \prod_{i=1}^{N} (x_i - t_i)^{\frac{\alpha_i}{k_i} - 1} f(t_1, ..., t_N)\, dt_1 ... dt_N, \tag{3.3.4}$$

where $\Gamma_{k_i}(\alpha_i)$ is the k_i-gamma function given by $\Gamma_{k_i}(\alpha_i) = \int_0^{\infty} t^{\alpha_i - 1} e^{-\frac{t^{k_i}}{k_i}} dt$, $i = 1, ..., N$ (it holds [22] $\Gamma_{k_i}(\alpha_i + k_i) = \alpha_i \Gamma_{k_i}(\alpha_i)$, $\Gamma(\alpha_i) = \lim_{k_i \to 1} \Gamma_{k_i}(\alpha_i)$, where Γ is the gamma function, and ${}_kI_{a+}^{0} f := f$), $f \in L_{\infty}\left(\prod_{i=1}^{N} [a_i, b_i]\right)$, $a = (a_1, ..., a_N)$, and $x = (x_1, ..., x_N) \in \prod_{i=1}^{N} [a_i, b_i]$.

By [6], we get that $\left({}_kI_{a+}^{\alpha} f\right)$ is a continuous function on $\prod_{i=1}^{N} [a_i, b_i]$. Furthermore by [6] we get that ${}_kI_{a+}^{\alpha}$ is a bounded linear operator, which is a positive operator.

We notice the following

$$\left|{}_kI_{a+}^{\alpha} f(x)\right| \leq$$

$$\frac{1}{\prod_{i=1}^{N} k_i \Gamma_{k_i}(\alpha_i)} \left(\int_{a_1}^{x_1} \cdots \int_{a_N}^{x_N} \prod_{i=1}^{N} (x_i - t_i)^{\frac{\alpha_i}{k_i} - 1} dt_1 ... dt_N\right) \|f\|_{\infty, \prod_{i=1}^{N} [a_i, b_i]} \tag{3.3.5}$$

$$= \frac{1}{\prod_{i=1}^{N} k_i \Gamma_{k_i}(\alpha_i)} \prod_{i=1}^{N} \frac{(x_i - a_i)^{\frac{\alpha_i}{k_i}}}{\left(\frac{\alpha_i}{k_i}\right)} \|f\|_{\infty, \prod_{i=1}^{N} [a_i, b_i]}$$

$$= \left(\prod_{i=1}^{N} \frac{(x_i - a_i)^{\frac{\alpha_i}{k_i}}}{\Gamma_{k_i}(\alpha_i + k_i)}\right) \|f\|_{\infty, \prod_{i=1}^{N} [a_i, b_i]}. \tag{3.3.6}$$

That is, it holds

$$\left|{}_kI_{a+}^{\alpha} f(x)\right| \leq \left(\prod_{i=1}^{N} \frac{(x_i - a_i)^{\frac{\alpha_i}{k_i}}}{\Gamma_{k_i}(\alpha_i + k_i)}\right) \|f\|_{\infty, \prod_{i=1}^{N} [a_i, b_i]}$$

$$\leq \left(\prod_{i=1}^{N} \frac{(b_i - a_i)^{\frac{\alpha_i}{k_i}}}{\Gamma_{k_i}(\alpha_i + k_i)}\right) \|f\|_{\infty, \prod_{i=1}^{N} [a_i, b_i]}. \tag{3.3.7}$$

We get that

$$ {}_kI^{\alpha}_{a+}f\left(a\right)=0. \tag{3.3.8}$$

In particular, $\left({}_kI^{\alpha}_{a+}f\right)$ is continuous on $\prod_{i=1}^{N}\left[a_i^*,b_i^*\right]$.

Thus there exist $x_1,x_2\in\prod_{i=1}^{N}\left[a_i^*,b_i^*\right]$ such that

$$\begin{aligned}\left({}_kI^{\alpha}_{a+}f\right)(x_1)&=\min\left({}_kI^{\alpha}_{a+}f\right)(x),\\ \left({}_kI^{\alpha}_{a+}f\right)(x_2)&=\max\left({}_kI^{\alpha}_{a+}f\right)(x),\end{aligned}\tag{3.3.9}$$

over all $x\in\prod_{i=1}^{N}\left[a_i^*,b_i^*\right]$.

We assume that

$$\left({}_kI^{\alpha}_{a+}f\right)(x_1)>0. \tag{3.3.10}$$

Hence

$$\left\|{}_kI^{\alpha}_{a+}f\right\|_{\infty,\prod_{i=1}^{N}\left[a_i^*,b_i^*\right]}=\left({}_kI^{\alpha}_{a+}f\right)(x_2)>0. \tag{3.3.11}$$

Here we define

$$Jf\left(x\right)=mf\left(x\right),\ 0<m<\frac{1}{2}, \tag{3.3.12}$$

for any $x\in\prod_{i=1}^{N}\left[a_i^*,b_i^*\right]$.

Therefore the equation

$$Jf\left(x\right)=0,\ x\in\prod_{i=1}^{N}\left[a_i^*,b_i^*\right], \tag{3.3.13}$$

has the same solutions as the equation

$$F\left(x\right):=\frac{Jf\left(x\right)}{2\left({}_kI^{\alpha}_{a+}f\right)(x_2)}=0,\ x\in\prod_{i=1}^{N}\left[a_i^*,b_i^*\right]. \tag{3.3.14}$$

Notice that

$${}_kI^{\alpha}_{a+}\left(\frac{f}{2\left({}_kI^{\alpha}_{a+}f\right)(x_2)}\right)(x)=\frac{\left({}_kI^{\alpha}_{a+}f\right)(x)}{2\left({}_kI^{\alpha}_{a+}f\right)(x_2)}\leq\frac{1}{2}<1,\ x\in\prod_{i=1}^{N}\left[a_i^*,b_i^*\right]. \tag{3.3.15}$$

Call

$$A(x) := \frac{\left({}_kI^{\alpha}_{a+}f\right)(x)}{2\left({}_kI^{\alpha}_{a+}f\right)(x_2)}, \ \forall\, x \in \prod_{i=1}^{N}\left[a_i^*, b_i^*\right]. \tag{3.3.16}$$

We notice that

$$0 < \frac{\left({}_kI^{\alpha}_{a+}f\right)(x_1)}{2\left({}_kI^{\alpha}_{a+}f\right)(x_2)} \leq A(x) \leq \frac{1}{2}, \ \forall\, x \in \prod_{i=1}^{N}\left[a_i^*, b_i^*\right]. \tag{3.3.17}$$

Hence the first condition (3.3.1) is fulfilled by

$$|1 - A(x)| = 1 - A(x) \leq 1 - \frac{\left({}_kI^{\alpha}_{a+}f\right)(x_1)}{2\left({}_kI^{\alpha}_{a+}f\right)(x_2)} =: \gamma_0, \ \forall\, x \in \prod_{i=1}^{N}\left[a_i^*, b_i^*\right]. \tag{3.3.18}$$

So that $\|1 - A(x)\|_\infty \leq \gamma_0$, where $\|\cdot\|_\infty$ is over $\prod_{i=1}^{N}\left[a_i^*, b_i^*\right]$. Clearly $\gamma_0 \in (0, 1)$.

Next we assume that $\frac{f(x)}{2\left({}_kI^{\alpha}_{a+}f\right)(x_2)}$ is a contraction, that is

$$\left|\frac{f(x)}{2\left({}_kI^{\alpha}_{a+}f\right)(x_2)} - \frac{f(y)}{2\left({}_kI^{\alpha}_{a+}f\right)(x_2)}\right| \leq \theta\,\|x - y\|, \ \text{all } x, y \in \prod_{i=1}^{N}\left[a_i^*, b_i^*\right], \tag{3.3.19}$$

$0 < \theta < 1$.

Hence

$$\left|\frac{mf(x)}{2\left({}_kI^{\alpha}_{a+}f\right)(x_2)} - \frac{mf(y)}{2\left({}_kI^{\alpha}_{a+}f\right)(x_2)}\right| \leq m\theta\,\|x - y\| \leq \frac{\theta}{2}\,\|x - y\|, \tag{3.3.20}$$

all $x, y \in \prod_{i=1}^{N}\left[a_i^*, b_i^*\right]$.

Set $\lambda = \frac{\theta}{2}$, it is $0 < \lambda < \frac{1}{2}$. We have that

$$|F(x) - F(y)| \leq \lambda\,\|x - y\|, \tag{3.3.21}$$

all $x, y \in \prod_{i=1}^{N}\left[a_i^*, b_i^*\right]$.

Equivalently we have

$$|Jf(x) - Jf(y)| \leq 2\lambda\left({}_kI^{\alpha}_{a+}f\right)(x_2)\,\|x - y\|, \ \text{all } x, y \in \prod_{i=1}^{N}\left[a_i^*, b_i^*\right]. \tag{3.3.22}$$

We observe that

$$\left\| \left(F\left(y\right) -F\left(x\right) \right) \overrightarrow{i}-A\left(x\right) \left(y-x\right) \right\| \leq$$

$$\left| F\left(y\right) -F\left(x\right) \right| +\left| A\left(x\right) \right| \left\| y-x\right\| \leq \tag{3.3.23}$$

$$\lambda \left\| y-x\right\| +\left| A\left(x\right) \right| \left\| y-x\right\| =\left(\lambda +\left| A\left(x\right) \right| \right) \left\| y-x\right\| =:\left(\psi_{1}\right) ,$$

$\forall\, x, y \in \prod_{i=1}^{N} \left[a_i^*, b_i^* \right]$.

By (3.3.7), we have that

$$\left| \left({}_{k}I_{a+}^{\alpha} f\right) \left(x\right) \right| \leq \left(\prod_{i=1}^{N} \frac{\left(b_i - a_i\right)^{\frac{\alpha_i}{k_i}}}{\Gamma_{k_i}\left(\alpha_i + k_i\right)} \right) \left\| f\right\|_{\infty}, \tag{3.3.24}$$

$\forall\, x \in \prod_{i=1}^{N} \left[a_i^*, b_i^* \right]$, where $\left\| \cdot \right\|_{\infty}$ now is over $\prod_{i=1}^{N} \left[a_i, b_i \right]$.

Hence

$$\left| A\left(x\right) \right| =\frac{\left| \left({}_{k}I_{a+}^{\alpha} f\right) \left(x\right) \right| }{2\left({}_{k}I_{a+}^{\alpha} f\right) \left(x_{2}\right) }\leq \frac{1}{2\left({}_{k}I_{a+}^{\alpha} f\right) \left(x_{2}\right) }\left(\prod_{i=1}^{N} \frac{\left(b_i - a_i\right)^{\frac{\alpha_i}{k_i}}}{\Gamma_{k_i}\left(\alpha_i + k_i\right)} \right) \left\| f\right\|_{\infty} <\infty , \tag{3.3.25}$$

$\forall\, x \in \prod_{i=1}^{N} \left[a_i^*, b_i^* \right]$.

Therefore we get

$$\left(\psi_{1}\right) \leq \left(\lambda +\frac{1}{2\left({}_{k}I_{a+}^{\alpha} f\right) \left(x_{2}\right) }\left(\prod_{i=1}^{N} \frac{\left(b_i - a_i\right)^{\frac{\alpha_i}{k_i}}}{\Gamma_{k_i}\left(\alpha_i + k_i\right)} \right) \left\| f\right\|_{\infty} \right) \left\| y-x\right\| , \tag{3.3.26}$$

$\forall\, x, y \in \prod_{i=1}^{N} \left[a_i^*, b_i^* \right]$.

Call

$$0<\gamma_{1}:=\lambda +\frac{1}{2\left({}_{k}I_{a+}^{\alpha} f\right) \left(x_{2}\right) }\left(\prod_{i=1}^{N} \frac{\left(b_i - a_i\right)^{\frac{\alpha_i}{k_i}}}{\Gamma_{k_i}\left(\alpha_i + k_i\right)} \right) \left\| f\right\|_{\infty} , \tag{3.3.27}$$

and by choosing $\left(b_i - a_i\right)$ small enough, $i = 1, ..., N$, we can make $\gamma_1 \in (0, 1)$, fulfilling (3.3.2).

Next we call and we need that

$$0<\gamma :=\gamma_{0}+\gamma_{1}=\left(1-\frac{\left({}_{k}I_{a+}^{\alpha} f\right) \left(x_{1}\right) }{2\left({}_{k}I_{a+}^{\alpha} f\right) \left(x_{2}\right) }\right) +$$

$$\left(\lambda + \frac{1}{2\left({}_kI^{\alpha}_{a+}f\right)(x_2)}\left(\prod_{i=1}^{N}\frac{(b_i-a_i)^{\frac{\alpha_i}{k_i}}}{\Gamma_{k_i}(\alpha_i+k_i)}\right)\|f\|_{\infty}\right) < 1, \tag{3.3.28}$$

equivalently,

$$\lambda + \frac{1}{2\left({}_kI^{\alpha}_{a+}f\right)(x_2)}\left(\prod_{i=1}^{N}\frac{(b_i-a_i)^{\frac{\alpha_i}{k_i}}}{\Gamma_{k_i}(\alpha_i+k_i)}\right)\|f\|_{\infty} < \frac{\left({}_kI^{\alpha}_{a+}f\right)(x_1)}{2\left({}_kI^{\alpha}_{a+}f\right)(x_2)}, \tag{3.3.29}$$

equivalently,

$$2\lambda\left({}_kI^{\alpha}_{a+}f\right)(x_2) + \left(\prod_{i=1}^{N}\frac{(b_i-a_i)^{\frac{\alpha_i}{k_i}}}{\Gamma_{k_i}(\alpha_i+k_i)}\right)\|f\|_{\infty} < \left({}_kI^{\alpha}_{a+}f\right)(x_1), \tag{3.3.30}$$

which is possible for small λ and small $(b_i - a_i)$, all $i = 1, ..., N$. That is $\gamma \in (0, 1)$, fulfilling (3.3.3). So our numerical method converges and solves (3.3.13).

(II) Consider the k-right multidimensional Riemann–Liouville fractional integral of order $\alpha = (\alpha_1, ..., \alpha_N)$, $k = (k_1, ..., k_N)$, $(\alpha_i > 0, k_i > 0, i = 1, ..., N)$:

$$\left({}_kI^{\alpha}_{b-}f\right)(x) = \frac{1}{\prod_{i=1}^{N}k_i\Gamma_{k_i}(\alpha_i)}\int_{x_1}^{b_1}...\int_{x_N}^{b_N}\prod_{i=1}^{N}(t_i-x_i)^{\frac{\alpha_i}{k_i}-1}f(t_1,...,t_N)\,dt_1...dt_N, \tag{3.3.31}$$

we set

$${}_kI^{0}_{b-}f = f, \tag{3.3.32}$$

where $f \in L_{\infty}\left(\prod_{i=1}^{N}[a_i, b_i]\right)$, $b = (b_1, ..., b_N)$, and $x = (x_1, ..., x_N) \in \prod_{i=1}^{N}[a_i, b_i]$.

By [7], we get that ${}_kI^{\alpha}_{b-}f$ is a continuous function on $\prod_{i=1}^{N}[a_i, b_i]$. Furthermore by [7] we get that ${}_kI^{\alpha}_{b-}$ is a bounded linear operator, which is a positive operator.

We notice the following

$$\left|{}_kI^{\alpha}_{b-}f(x)\right| \leq \frac{1}{\prod_{i=1}^{N}k_i\Gamma_{k_i}(\alpha_i)}\prod_{i=1}^{N}\frac{(b_i-x_i)^{\frac{\alpha_i}{k_i}}}{\left(\frac{\alpha_i}{k_i}\right)}\|f\|_{\infty}$$

$$= \left(\prod_{i=1}^{N}\frac{(b_i-x_i)^{\frac{\alpha_i}{k_i}}}{\Gamma_{k_i}(\alpha_i+k_i)}\right)\|f\|_{\infty}. \tag{3.3.33}$$

That is it holds

$$\left|{}_kI_{b-}^{\alpha} f(x)\right| \leq \left(\prod_{i=1}^{N} \frac{(b_i - x_i)^{\frac{\alpha_i}{k_i}}}{\Gamma_{k_i}(\alpha_i + k_i)}\right) \|f\|_\infty$$

$$\leq \left(\prod_{i=1}^{N} \frac{(b_i - a_i)^{\frac{\alpha_i}{k_i}}}{\Gamma_{k_i}(\alpha_i + k_i)}\right) \|f\|_\infty . \tag{3.3.34}$$

We get that

$${}_kI_{b-}^{\alpha} f(b) = 0. \tag{3.3.35}$$

In particular, $\left({}_kI_{b-}^{\alpha} f\right)$ is continuous on $\prod_{i=1}^{N} \left[a_i^*, b_i^*\right]$.

Thus there exist $x_1, x_2 \in \prod_{i=1}^{N} \left[a_i^*, b_i^*\right]$ such that

$$\begin{aligned} \left({}_kI_{b-}^{\alpha} f\right)(x_1) &= \min \left({}_kI_{b-}^{\alpha} f\right)(x), \\ \left({}_kI_{b-}^{\alpha} f\right)(x_2) &= \max \left({}_kI_{b-}^{\alpha} f\right)(x), \end{aligned} \tag{3.3.36}$$

over all $x \in \prod_{i=1}^{N} \left[a_i^*, b_i^*\right]$.

We assume that

$$\left({}_kI_{b-}^{\alpha} f\right)(x_1) > 0. \tag{3.3.37}$$

Hence

$$\left\|{}_kI_{b-}^{\alpha} f\right\|_{\infty, \prod_{i=1}^{N}\left[a_i^*, b_i^*\right]} = \left({}_kI_{b-}^{\alpha} f\right)(x_2) > 0. \tag{3.3.38}$$

Here we define

$$Jf(x) = mf(x), 0 < m < \frac{1}{2}, \tag{3.3.39}$$

for any $x \in \prod_{i=1}^{N} \left[a_i^*, b_i^*\right]$.

Therefore the equation

$$Jf(x) = 0, x \in \prod_{i=1}^{N} \left[a_i^*, b_i^*\right], \tag{3.3.40}$$

has the same solutions as the equation

$$F(x) := \frac{Jf(x)}{2\left({}_kI_{b-}^{\alpha} f\right)(x_2)} = 0, \ x \in \prod_{i=1}^{N} \left[a_i^*, b_i^*\right]. \tag{3.3.41}$$

Notice that

$$ {}_kI^{\alpha}_{b-}\left(\frac{f}{2\left({}_kI^{\alpha}_{b-}f\right)(x_2)}\right)(x)=\frac{\left({}_kI^{\alpha}_{b-}f\right)(x)}{2\left({}_kI^{\alpha}_{b-}f\right)(x_2)}\leq\frac{1}{2}<1,\ x\in\prod_{i=1}^{N}\left[a_i^*,b_i^*\right]. \tag{3.3.42}$$

Call

$$ A(x):=\frac{\left({}_kI^{\alpha}_{b-}f\right)(x)}{2\left({}_kI^{\alpha}_{b-}f\right)(x_2)},\ \forall\, x\in\prod_{i=1}^{N}\left[a_i^*,b_i^*\right]. \tag{3.3.43}$$

We notice that

$$ 0<\frac{\left({}_kI^{\alpha}_{b-}f\right)(x_1)}{2\left({}_kI^{\alpha}_{b-}f\right)(x_2)}\leq A(x)\leq\frac{1}{2},\ \forall\, x\in\prod_{i=1}^{N}\left[a_i^*,b_i^*\right]. \tag{3.3.44}$$

Hence the first condition (3.3.1) is fulfilled by

$$ |1-A(x)|=1-A(x)\leq 1-\frac{\left({}_kI^{\alpha}_{b-}f\right)(x_1)}{2\left({}_kI^{\alpha}_{b-}f\right)(x_2)}=:\gamma_0,\ \forall\, x\in\prod_{i=1}^{N}\left[a_i^*,b_i^*\right]. \tag{3.3.45}$$

So that $\|1-A(x)\|_{\infty}\leq\gamma_0$, where $\|\cdot\|_{\infty}$ is over $\prod_{i=1}^{N}\left[a_i^*,b_i^*\right]$. Clearly $\gamma_0\in(0,1)$.

Next we assume that $\frac{f(x)}{2\left({}_kI^{\alpha}_{b-}f\right)(x_2)}$ is a contraction, that is

$$ \left|\frac{f(x)}{2\left({}_kI^{\alpha}_{b-}f\right)(x_2)}-\frac{f(y)}{2\left({}_kI^{\alpha}_{b-}f\right)(x_2)}\right|\leq\theta\|x-y\|,\ \text{all}\ x,y\in\prod_{i=1}^{N}\left[a_i^*,b_i^*\right], \tag{3.3.46}$$

$0<\theta<1$.

Hence

$$ \left|\frac{mf(x)}{2\left({}_kI^{\alpha}_{b-}f\right)(x_2)}-\frac{mf(y)}{2\left({}_kI^{\alpha}_{b-}f\right)(x_2)}\right|\leq m\theta\|x-y\|\leq\frac{\theta}{2}\|x-y\|, \tag{3.3.47}$$

all $x,y\in\prod_{i=1}^{N}\left[a_i^*,b_i^*\right]$.

Set $\lambda=\frac{\theta}{2}$, it is $0<\lambda<\frac{1}{2}$. We have that

$$ |F(x)-F(y)|\leq\lambda\|x-y\|, \tag{3.3.48}$$

all $x,y\in\prod_{i=1}^{N}\left[a_i^*,b_i^*\right]$.

Equivalently we have

$$|Jf(x) - Jf(y)| \leq 2\lambda \left({}_kI_{b-}^{\alpha} f\right)(x_2) \|x - y\|, \text{ all } x, y \in \prod_{i=1}^{N} \left[a_i^*, b_i^*\right]. \quad (3.3.49)$$

We observe that

$$\left\| (F(y) - F(x)) \overrightarrow{i} - A(x)(y - x) \right\| \leq$$

$$|F(y) - F(x)| + |A(x)| \|y - x\| \leq \quad (3.3.50)$$

$$\lambda \|y - x\| + |A(x)| \|y - x\| = (\lambda + |A(x)|) \|y - x\| =: (\psi_2),$$

$\forall\, x, y \in \prod_{i=1}^{N} \left[a_i^*, b_i^*\right]$.

By (3.3.34), we have that

$$\left| \left({}_kI_{b-}^{\alpha} f\right)(x) \right| \leq \left(\prod_{i=1}^{N} \frac{(b_i - a_i)^{\frac{\alpha_i}{k_i}}}{\Gamma_{k_i}(\alpha_i + k_i)} \right) \|f\|_{\infty}, \quad (3.3.51)$$

$\forall\, x \in \prod_{i=1}^{N} \left[a_i^*, b_i^*\right]$, where $\|\cdot\|_{\infty}$ now is over $\prod_{i=1}^{N} [a_i, b_i]$.

Hence

$$|A(x)| = \frac{\left| \left({}_kI_{b-}^{\alpha} f\right)(x) \right|}{2\left({}_kI_{b-}^{\alpha} f\right)(x_2)} \leq \frac{1}{2\left({}_kI_{b-}^{\alpha} f\right)(x_2)} \left(\prod_{i=1}^{N} \frac{(b_i - a_i)^{\frac{\alpha_i}{k_i}}}{\Gamma_{k_i}(\alpha_i + k_i)} \right) \|f\|_{\infty} < \infty, \quad (3.3.52)$$

$\forall\, x \in \prod_{i=1}^{N} \left[a_i^*, b_i^*\right]$.

Therefore we get

$$(\psi_2) \leq \left(\lambda + \frac{1}{2\left({}_kI_{b-}^{\alpha} f\right)(x_2)} \left(\prod_{i=1}^{N} \frac{(b_i - a_i)^{\frac{\alpha_i}{k_i}}}{\Gamma_{k_i}(\alpha_i + k_i)} \right) \|f\|_{\infty} \right) \|y - x\|, \quad (3.3.53)$$

$\forall\, x, y \in \prod_{i=1}^{N} \left[a_i^*, b_i^*\right]$.

Call

$$0 < \gamma_1 := \lambda + \frac{1}{2\left({}_kI_{b-}^{\alpha} f\right)(x_2)} \left(\prod_{i=1}^{N} \frac{(b_i - a_i)^{\frac{\alpha_i}{k_i}}}{\Gamma_{k_i}(\alpha_i + k_i)} \right) \|f\|_{\infty}, \quad (3.3.54)$$

and by choosing $(b_i - a_i)$ small enough, $i = 1, ..., N$, we can make $\gamma_1 \in (0, 1)$, fulfilling (3.3.2).

Next we call and we need that

$$0 < \gamma := \gamma_0 + \gamma_1 = \left(1 - \frac{\left({}_kI_{b-}^{\alpha} f\right)(x_1)}{2\left({}_kI_{b-}^{\alpha} f\right)(x_2)}\right) +$$

$$\left(\lambda + \frac{1}{2\left({}_kI_{b-}^{\alpha} f\right)(x_2)}\left(\prod_{i=1}^{N} \frac{(b_i - a_i)^{\frac{\alpha_i}{k_i}}}{\Gamma_{k_i}(\alpha_i + k_i)}\right)\|f\|_{\infty}\right) < 1, \tag{3.3.55}$$

equivalently,

$$\lambda + \frac{1}{2\left({}_kI_{b-}^{\alpha} f\right)(x_2)}\left(\prod_{i=1}^{N} \frac{(b_i - a_i)^{\frac{\alpha_i}{k_i}}}{\Gamma_{k_i}(\alpha_i + k_i)}\right)\|f\|_{\infty} < \frac{\left({}_kI_{b-}^{\alpha} f\right)(x_1)}{2\left({}_kI_{b-}^{\alpha} f\right)(x_2)}, \tag{3.3.56}$$

equivalently,

$$2\lambda\left({}_kI_{b-}^{\alpha} f\right)(x_2) + \left(\prod_{i=1}^{N} \frac{(b_i - a_i)^{\frac{\alpha_i}{k_i}}}{\Gamma_{k_i}(\alpha_i + k_i)}\right)\|f\|_{\infty} < \left({}_kI_{b-}^{\alpha} f\right)(x_1), \tag{3.3.57}$$

which is possible for small λ and small $(b_i - a_i)$, all $i = 1, ..., N$. That is $\gamma \in (0, 1)$, fulfilling (3.3.3). So our numerical method converges and solves (3.3.40).

(III) Here we deal with the following multivariate mixed fractional derivative: let $\alpha = (\alpha_1, ..., \alpha_N)$, where $0 < \alpha_i < 1$, $i = 1, ..., N$; $f \in C^N\left(\prod_{i=1}^{N}[0, b_i]\right)$; $b_i > 0$, $i = 1, ..., N$,

$$\left({}^{CF}D_*^{\alpha} f\right)(t) = \frac{1}{\prod_{i=1}^{N}(1 - \alpha_i)}\cdot \tag{3.3.58}$$

$$\int_0^{t_1} \cdots \int_0^{t_N} \prod_{i=1}^{N} \exp\left(-\frac{\alpha_i}{1 - \alpha_i}(t_i - s_i)\right) \frac{\partial^N f(s_1, ..., s_N)}{\partial s_1 ... \partial s_N} ds_1 ... ds_N,$$

for all $0 \le t_i \le b_i$, $i = 1, ..., N$; with $t = (t_1, ..., t_N)$.

When $N = 1$, the univariate case is known as the M. Caputo-Fabrizio fractional derivative, see [19].

Call

$$\gamma_i := \frac{\alpha_i}{1 - \alpha_i} > 0. \tag{3.3.59}$$

I.e.

$$\left({}^{CF}D_*^{\alpha} f\right)(t) = \frac{1}{\prod_{i=1}^{N}(1 - \alpha_i)}\cdot \tag{3.3.60}$$

$$\int_0^{t_1} \dots \int_0^{t_N} \prod_{i=1}^{N} e^{-\gamma_i (t_i - s_i)} \frac{\partial^N f(s_1, \dots, s_N)}{\partial s_1 \dots \partial s_N} ds_1 \dots ds_N,$$

all $0 \leq t_i \leq b_i$, $i = 1, \dots, N$.

We notice that

$$\left| \left({}^{CF}D_*^{\alpha} f \right)(t) \right| \leq \frac{1}{\prod_{i=1}^{N} (1 - \alpha_i)} \cdot$$

$$\left(\int_0^{t_1} \dots \int_0^{t_N} \prod_{i=1}^{N} e^{-\gamma_i (t_i - s_i)} ds_1 \dots ds_N \right) \left\| \frac{\partial^N f}{\partial x_1 \dots \partial x_N} \right\|_{\infty} = \tag{3.3.61}$$

$$\prod_{i=1}^{N} \left(\frac{1}{1 - \alpha_i} \int_0^{t_i} e^{-\gamma_i (t_i - s_i)} ds_i \right) \left\| \frac{\partial^N f}{\partial x_1 \dots \partial x_N} \right\|_{\infty} =$$

$$\left(\prod_{i=1}^{N} \frac{e^{-\gamma_i t_i}}{\alpha_i} \left(e^{\gamma_i t_i} - 1 \right) \right) \left\| \frac{\partial^N f}{\partial x_1 \dots \partial x_N} \right\|_{\infty} =$$

$$\left(\prod_{i=1}^{N} \frac{1}{\alpha_i} \left(1 - e^{-\gamma_i t_i} \right) \right) \left\| \frac{\partial^N f}{\partial x_1 \dots \partial x_N} \right\|_{\infty} \leq$$

$$\left(\prod_{i=1}^{N} \left(\frac{1 - e^{-\gamma_i b_i}}{\alpha_i} \right) \right) \left\| \frac{\partial^N f}{\partial x_1 \dots \partial x_N} \right\|_{\infty} . \tag{3.3.62}$$

That is

$$\left({}^{CF}D_*^{\alpha} f \right)(0, \dots, 0) = 0, \tag{3.3.63}$$

and

$$\left| \left({}^{CF}D_*^{\alpha} f \right)(t) \right| \leq \left(\prod_{i=1}^{N} \left(\frac{1 - e^{-\gamma_i b_i}}{\alpha_i} \right) \right) \left\| \frac{\partial^N f}{\partial x_1 \dots \partial x_N} \right\|_{\infty} . \tag{3.3.64}$$

Notice here that $1 - e^{-\gamma_i t_i}$, $t_i \geq 0$ is an increasing function, $i = 1, \dots, N$.

Thus the smaller the t_i, the smaller it is $1 - e^{-\gamma_i t_i}$, $i = 1, \dots, N$. We can rewrite

$$\left({}^{CF}D_*^{\alpha} f \right)(t) =$$

$$\prod_{i=1}^{N} \left(\frac{e^{-\gamma_i t_i}}{1 - \alpha_i} \right) \int_0^{t_1} \dots \int_0^{t_N} e^{\sum_{i=1}^{N} \gamma_i s_i} \frac{\partial^N f(s_1, \dots, s_N)}{\partial s_1 \dots \partial s_N} ds_1 \dots ds_N = \tag{3.3.65}$$

$$\prod_{i=1}^{N}\left(\frac{e^{-\gamma_i t_i}}{1-\alpha_i}\right).$$

$$\int_0^{b_1}...\int_0^{b_N}\chi_{\prod_{i=1}^{N}[0,t_i]}(s_1,...,s_N)\,e^{\sum_{i=1}^{N}\gamma_i s_i}\frac{\partial^N f(s_1,...,s_N)}{\partial s_1...\partial s_N}ds_1...ds_N, \quad (3.3.66)$$

where χ stands for the characteristic function.

Let $t_n \to t$, as $n \to \infty$, then $\chi_{\prod_{i=1}^{N}[0,t_{in}]}(s_1,...,s_N) \to \chi_{\prod_{i=1}^{N}[0,t_i]}(s_1,...,s_N)$, a.e, as $n \to \infty$, where $t_n = (t_{1n},...,t_{Nn})$.

Hence we have $\chi_{\prod_{i=1}^{N}[0,t_{in}]}(s_1,...,s_N)\,e^{\sum_{i=1}^{N}\gamma_i s_i}\frac{\partial^N f(s_1,...,s_N)}{\partial s_1...\partial s_N} \to$ $\chi_{\prod_{i=1}^{N}[0,t_i]}(s_1,...,s_N)\,e^{\sum_{i=1}^{N}\gamma_i s_i}\frac{\partial^N f(s_1,...,s_N)}{\partial s_1...\partial s_N}$, a.e., in $(s_1,...,s_N) \in \prod_{i=1}^{N}[0,b_i]$.

Furthermore, it holds

$$\chi_{\prod_{i=1}^{N}[0,t_{iN}]}(s_1,...,s_N)\,e^{\sum_{i=1}^{N}\gamma_i s_i}\left|\frac{\partial^N f(s_1,...,s_N)}{\partial s_1...\partial s_N}\right|$$

$$\leq e^{\sum_{i=1}^{N}\gamma_i b_i}\left\|\frac{\partial^N f}{\partial x_1...\partial x_N}\right\|_{\infty}. \quad (3.3.67)$$

Thus, by dominated convergence theorem we get

$$\left({}^{CF}D_*^{\alpha}f\right)(t_n) \to \left({}^{CF}D_*^{\alpha}f\right)(t), \text{ as } n \to \infty,$$

proving continuity of $\left({}^{CF}D_*^{\alpha}f\right)(t)$, $t \in \prod_{i=1}^{N}[0,b_i]$.

In particular, $\left({}^{CF}D_*^{\alpha}f\right)(t)$ is continuous, $\forall\, t \in \prod_{i=1}^{N}[a_i,b_i]$, where $0 < a_i < b_i$, $i = 1,...,N$.

Therefore there exist $x_1, x_2 \in \prod_{i=1}^{N}[a_i,b_i]$ such that

$${}^{CF}D_*^{\alpha}f(x_1) = \min\ {}^{CF}D_*^{\alpha}f(x), \quad (3.3.68)$$

and

$${}^{CF}D_*^{\alpha}f(x_2) = \max\ {}^{CF}D_*^{\alpha}f(x), \text{ for } x \in \prod_{i=1}^{N}[a_i,b_i]. \quad (3.3.69)$$

We assume that

$${}^{CF}D_*^{\alpha}f(x_1) > 0. \quad (3.3.70)$$

(i.e. ${}^{CF}D_*^{\alpha}f(x) > 0$, $\forall\, x \in \prod_{i=1}^{N}[a_i,b_i]$).

Furthermore

$$\left\|{}^{CF}D_*^{\alpha}fG\right\|_{\infty,[a,b]} = {}^{CF}D_*^{\alpha}f(x_2). \quad (3.3.71)$$

Here we define

$$J f(x) = m f(x), 0 < m < \frac{1}{2}, \forall x \in \prod_{i=1}^{N} [a_i, b_i]. \tag{3.3.72}$$

The equation

$$J f(x) = 0, \ x \in \prod_{i=1}^{N} [a_i, b_i], \tag{3.3.73}$$

has the same set of solutions as the equation

$$F(x) := \frac{J f(x)}{{}^{CF}D_*^{\alpha} f(x_2)} = 0, x \in \prod_{i=1}^{N} [a_i, b_i]. \tag{3.3.74}$$

Notice that

$${}^{CF}D_*^{\alpha} \left(\frac{f(x)}{2\,{}^{CF}D_*^{\alpha} f(x_2)} \right) = \frac{{}^{CF}D_*^{\alpha} f(x)}{2\,{}^{CF}D_*^{\alpha} f(x_2)} \leq \frac{1}{2} < 1, \ \forall x \in \prod_{i=1}^{N} [a_i, b_i]. \tag{3.3.75}$$

We call

$$A(x) := \frac{{}^{CF}D_*^{\alpha} f(x)}{2\,{}^{CF}D_*^{\alpha} f(x_2)}, \forall x \in \prod_{i=1}^{N} [a_i, b_i]. \tag{3.3.76}$$

We notice that

$$0 < \frac{{}^{CF}D_*^{\alpha} f(x_1)}{2\,{}^{CF}D_*^{\alpha} f(x_2)} \leq A(x) \leq \frac{1}{2}. \tag{3.3.77}$$

Furthermore it holds

$$|1 - A(x)| = 1 - A(x) \leq 1 - \frac{{}^{CF}D_*^{\alpha} f(x_1)}{2\,{}^{CF}D_*^{\alpha} f(x_2)} =: \gamma_0, \ \forall x \in \prod_{i=1}^{N} [a_i, b_i]. \tag{3.3.78}$$

Clearly $\gamma_0 \in (0, 1)$.

We have proved that

$$\|1 - A(x)\|_{\infty} \leq \gamma_0 \in (0, 1), \ \forall x \in \prod_{i=1}^{N} [a_i, b_i], \tag{3.3.79}$$

see (3.3.1) fulfilled.

Next we assume that $F(x)$ is a contraction over $\prod_{i=1}^{N}[a_i, b_i]$, i.e.

$$|F(x) - F(y)| \leq \lambda \|x - y\|; \ \forall x, y \in \prod_{i=1}^{N}[a_i, b_i], \tag{3.3.80}$$

and $0 < \lambda < \frac{1}{2}$.

Equivalently we have

$$|Jf(x) - Jf(y)| \leq 2\lambda \left({}^{CF}D_*^{\alpha} f(x_2)\right) \|x - y\|, \ \forall x, y \in [a, b]. \tag{3.3.81}$$

We observe that

$$\left\|(F(y) - F(x))\overrightarrow{i} - A(x)(y - x)\right\| \leq |F(y) - F(x)| + |A(x)| \|y - x\| \leq$$

$$\lambda \|y - x\| + |A(x)| \|y - x\| = (\lambda + |A(x)|) \|y - x\| =: (\xi), \forall x, y \in \prod_{i=1}^{N}[a_i, b_i], \tag{3.3.82}$$

where $\overrightarrow{i}$ the unit vector in $\mathbb{R}^N$.

Here we have (3.3.64) valid on $\prod_{i=1}^{N}[a_i, b_i]$.

Hence, $\forall x \in \prod_{i=1}^{N}[a_i, b_i]$ we get that

$$|A(x)| = \frac{\left|{}^{CF}D_*^{\alpha} f(x)\right|}{2\left({}^{CF}D_*^{\alpha} f\right)(x_2)} \leq \frac{\left(\prod_{i=1}^{N}\left(\frac{1-e^{-\gamma_i b_i}}{\alpha_i}\right)\right)\left\|\frac{\partial^N f}{\partial x_1 \ldots \partial x_N}\right\|_{\infty}}{2\alpha\left({}^{CF}D_*^{\alpha} f\right)(x_2)} < \infty. \tag{3.3.83}$$

Consequently we observe

$$(\xi) \leq \left(\lambda + \frac{\left(\prod_{i=1}^{N}\left(\frac{1-e^{-\gamma_i b_i}}{\alpha_i}\right)\right)\left\|\frac{\partial^N f}{\partial x_1 \ldots \partial x_N}\right\|_{\infty}}{2\alpha\left({}^{CF}D_*^{\alpha} f\right)(x_2)}\right) \|y - x\|, \ \forall x, y \in \prod_{i=1}^{N}[a_i, b_i]. \tag{3.3.84}$$

Call

$$0 < \gamma_1 := \lambda + \frac{\left(\prod_{i=1}^{N}\left(\frac{1-e^{-\gamma_i b_i}}{\alpha_i}\right)\right)\left\|\frac{\partial^N f}{\partial x_1 \ldots \partial x_N}\right\|_{\infty}}{2\alpha\left({}^{CF}D_*^{\alpha} f\right)(x_2)}, \tag{3.3.85}$$

choosing b_i small enough, $i = 1, \ldots, N$, we can make $\gamma_1 \in (0, 1)$.

We have proved (3.3.2) over $\prod_{i=1}^{N}[a_i, b_i]$.

Next we call and need

$$0 < \gamma := \gamma_0 + \gamma_1 =$$

$$1 - \frac{{}^{CF}D_*^{\alpha} f(x_1)}{2^{CF}D_*^{\alpha} f(x_2)} + \lambda + \frac{\left(\prod_{i=1}^{N}\left(\frac{1-e^{-\gamma_i b_i}}{\alpha_i}\right)\right)\left\|\frac{\partial^N f}{\partial x_1 \ldots \partial x_N}\right\|_{\infty}}{2\alpha\left({}^{CF}D_*^{\alpha} f\right)(x_2)} < 1, \tag{3.3.86}$$

equivalently,

$$\lambda + \frac{\left(\prod_{i=1}^{N}\left(\frac{1-e^{-\gamma_i b_i}}{\alpha_i}\right)\right)\left\|\frac{\partial^N f}{\partial x_1 \ldots \partial x_N}\right\|_\infty}{2\alpha\left({}^{CF}D_*^\alpha f\right)(x_2)} < \frac{{}^{CF}D_*^\alpha f(x_1)}{2\,{}^{CF}D_*^\alpha f(x_2)}, \tag{3.3.87}$$

equivalently,

$$2\lambda\,{}^{CF}D_*^\alpha f(x_2) + \left(\prod_{i=1}^{N}\left(\frac{1-e^{-\gamma_i b_i}}{\alpha_i}\right)\right)\left\|\frac{\partial^N f}{\partial x_1 \ldots \partial x_N}\right\|_\infty < {}^{CF}D_*^\alpha f(x_1), \tag{3.3.88}$$

which is possible for small λ, b_i, $i = 1, ..., N$.

We have proved that

$$\gamma = \gamma_0 + \gamma_1 \in (0, 1), \tag{3.3.89}$$

fulfilling (3.3.3).

Hence Eq. (3.3.73) can be solved with our presented numerical methods.

Consequently, our presented Numerical methods here, Theorem 3.1, apply to solve

$$f(x) = 0. \tag{3.3.90}$$

References

1. S. Amat, S. Busquier, S. Plaza, Chaotic dynamics of a third-order Newton-type method. J. Math. Anal. Appl. **366**(1), 164–174 (2010)
2. G. Anastassiou, *Fractional Differentiation Inequalities* (Springer, New York, 2009)
3. G. Anastassiou, *Intelligent Mathematics: Computational Analysis* (Springer, Heidelberg, 2011)
4. G. Anastassiou, Fractional representation formulae and right fractional inequalities. Math. Comput. Model. **54**(10–12), 3098–3115 (2011)
5. G. Anastassiou, *Advanced Inequalities* (World Scientific Publishing Corporation, Singapore, 2011)
6. G. Anastassiou, On left multidimensional Riemann-Liouville fractional integral. J. Comput. Anal. Appl. (accepted, 2015)
7. G. Anastassiou, On right multidimensional Riemann-Liouville fractional integral. J. Comput. Anal. Appl. (accepted, 2015)
8. G. Anastassiou, I. Argyros, Results on the semi-local convergence of iterative methods with applications in k-multivariate fractional calculus (submitted, 2015)
9. G. Anastassiou, I.K. Argyros, Studies in computational intelligence, *Intelligent Numerical Methods: Applications to Fractional Calculus*, vol. 624 (Springer, Heidelberg, 2016)
10. I.K. Argyros, A unifying local-semilocal convergence analysis and applications for two-point Newton-like methods in Banach space. J. Math. Anal. Appl. **298**, 374–397 (2004)
11. I.K. Argyros, *Convergence and Applications of Newton-Type Iterations* (Springer, New York, 2008)
12. I.K. Argyros, On a class of Newton-like methods for solving nonlinear equations. J. Comput. Appl. Math. **228**, 115–122 (2009)
13. I.K. Argyros, A semilocal convergence analysis for directional Newton methods. Math. Comput. AMS **80**, 327–343 (2011)

14. I.K. Argyros, S. Hilout, Weaker conditions for the convergence of Newton's method. J. Complex. **28**, 364–387 (2012)
15. I.K. Argyros, Y.J. Cho, S. Hilout, *Numerical Methods for Equations and Its Applications* (CRC Press/Taylor and Francis Group, New York, 2012)
16. J.A. Ezquérro, M.A. Hernández, Newton-type methods of high order and domains of semilocal and global convergence. Appl. Math. Comput. **214**(1), 142–154 (2009)
17. J.A. Ezquérro, J.M. Gutiérrez, M.A. Hernández, N. Romero, M.J. Rubio, The Newton method: from Newton to Kantorovich (Spanish). Gac. R. Soc. Mat. Esp. **13**, 53–76 (2010)
18. L.V. Kantorovich, G.P. Akilov, *Functional Analysis in Normed Spaces* (Pergamon Press, New York, 1964)
19. J. Losada, J.J. Nieto, Properties of a new fractional derivative without singular Kernel. Prog. Fract. Differ. Appl. **1**(2), 87–92 (2015)
20. A.A. Magreñán, Different anomalies in a Jarratt family of iterative root finding methods. Appl. Math. Comput. **233**, 29–38 (2014)
21. A.A. Magreñán, A new tool to study real dynamics: the convergence plane. Appl. Math. Comput. **248**, 215–224 (2014)
22. S. Mubeen, G.M. Habibullah, *k*-fractional integrals and applications. Int. J. Contemp. Math. Sci. **7**(2), 89–94 (2012)

Chapter 4
Newton-Like Procedures and Their Applications in Multivariate Fractional Calculus

Newton-like methods are used to generate sequences converging to a solution of a nonlinear equation defined between two Banach spaces. In this paper a semi-local convergence analysis is presented for these methods under very general Newton-Mysovskii-type conditions suitable for a wider range of problems than in earlier studies. In particular, we suggest some applications in multivariate fractional calculus. It follows [7].

4.1 Introduction

Let B_1, B_2 denote Banach spaces and $\mathcal{L}(B_1, B_2)$ stand for the space of bounded linear operators from B_1 into B_2. The notations $U(y, \tau)$, $\overline{U}(y, \tau)$ will be used for the open and closed balls in B_1, respectively with center $y \in B_1$ and of radius $\tau > 0$.

On of the greatest challenges in Computational Mathematics is to approximate a solution υ^* of the equation

$$H(x) = 0, \tag{4.1.1}$$

where $H : \Omega \subseteq B_1 \to B_2$ and Ω is a subset of B_1.

Notice a lot of problems can be formulated as a special case of Eq. (4.1.1) using Mathematical Modelling [1–12, 17, 18].

The solutions υ^* can rarely be found in closed form. That is why we most solution methods for these equations are usually iterative.

We consider the Newton-like method defined for each $n = 0, 1, 2, ...$ by

$$\upsilon_{n+1} = \upsilon_n - A(\upsilon_n)^{-1} H(\upsilon_n), \tag{4.1.2}$$

where $\upsilon_0 \in \Omega$ is an initial guess and $A(\cdot) : \Omega \to \mathcal{L}(B_1, B_2)$. Method (4.1.2) specializes to many popular iterative schemes: if $A(x) = H'(x)$ for each $x \in \Omega$, we obtain Newton's method, where as, if $A(x) = [x, g(x); H]$, where $g : B_1 \to B_1$ is

G.A. Anastassiou and I.K. Argyros, *Intelligent Numerical Methods II: Applications to Multivariate Fractional Calculus*, Studies in Computational Intelligence 649, DOI 10.1007/978-3-319-33606-0_4

a known continuous operator and $[v, w; H]$ is a divided difference of order one for operator H, then we obtain the Steffensen's method [9].

Other choices for A as well as a survey of local and semi-local convergence results for iterative methods can be found in [1–18] and the references there in.

In this paper we present a semi-local convergence analysis for method (4.1.2) under generalized Newton–Mysovskii-type [9, 17], conditions which are suitable for a wider range of problems than before such as problems from multivariate fractional calculus [2–6] which are very important in many areas.

The rest of the chapter is organized as follows: in Sect. 4.2 we present the semi-local convergence of method (4.1.2). Some applications in multivariate fractional calculus are suggested in the concluding Sect. 4.3.

4.2 Semi-local Convergence

Some very general semi-local convergence results are presented in this section suitable for many applications, including applications to multivariate fractional calculus.

Let $R_0 > 0$. Define $R = \sup\{t \in [0, R_0] : U(x_0, t) \subseteq \Omega\}$. Set $\Omega_0 = \overline{U}(v_0, R)$.

Theorem 4.1 *Let $H : \Omega \subseteq B_1 \to B_2$ be a Fréchet-differentiable operator and let $A(\cdot) : \Omega_0 \to \mathcal{L}(B_1, B_2)$ be a continuous operator. Suppose that there exist $\xi \geq 0$, $s \geq 0$, $\gamma \geq 0$ and $\delta \geq 0$ such that*

$$A(x)^{-1} \in \mathcal{L}(B_2, B_1) \text{ for each } x \in \Omega_0 \tag{4.2.1}$$

$$\left\| A(v_0)^{-1} H(v_0) \right\| \leq \xi, \tag{4.2.2}$$

$$\left\| A(y)^{-1} \left(H(y) - H(x) - H'(x)(y - x) \right) \right\| \leq \gamma \|y - x\|^{s+1} \tag{4.2.3}$$

for each $x, y \in \Omega_1 := \Omega \cap \Omega_0$

$$\left\| A(y)^{-1} \left(A(x) - H'(x) \right) \right\| \leq \delta \text{ for each } x \in \Omega_1, \tag{4.2.4}$$

$$\gamma \xi^s + \delta < 1 \tag{4.2.5}$$

and

$$\overline{U}(v_0, \rho) \subseteq \Omega_0, \tag{4.2.6}$$

where

$$\rho = \frac{\xi}{1 - (\gamma \xi^s + \delta)}. \tag{4.2.7}$$

Then, sequence $\{v_n\}$ generated for $v_0 \in \Omega$ by method (4.1.2) is well defined, remains in $\overline{U}(v_0, \rho)$ for each $n = 0, 1, 2, \ldots$ and converges to some $v^* \in \overline{U}(v_0, \rho)$. Moreover the following error bounds hold

$$\|v_{n+1} - v_n\| \leq \left(\gamma \|v_n - v_{n-1}\|^s + \delta\right) \|v_n - v_{n-1}\| \leq r \|v_n - v_{n-1}\| \tag{4.2.8}$$

and

$$\left\|v_n - v^*\right\| \leq \frac{r^n \xi}{1 - r}, \tag{4.2.9}$$

where

$$r := \gamma \xi^s + \delta. \tag{4.2.10}$$

Proof Notice that in view of (4.2.5) and (4.2.7), $r \in [0, 1)$ and $\rho \geq 0$. By hypothesis (4.2.1) for $x = v_0$ iterative method (4.1.2) for $n = 0$, point v_1 exists. Then, by (4.2.2) and (4.2.7), we have that

$$\|v_1 - v_0\| = \left\|A(v_0)^{-1} H(v_0)\right\| \leq \xi \leq \rho,$$

so $v_1 \in \overline{U}(v_0, \rho)$ and v_2 exists. Using (4.2.3) and (4.2.4), we get in turn that

$$\begin{aligned}
\|v_2 - v_1\| &= \left\|A(v_1)^{-1}(H(v_1) - H(v_0) - A(v_0)(v_1 - v_0))\right\| \\
&\leq \left\|A(v_1)^{-1}\left(H(v_1) - H(v_0) - H'(v_0)(v_1 - v_0)\right)\right\| \\
&\quad + \left\|A(v_1)^{-1}\left(A(v_0) - H'(v_0)\right)\right\| \\
&\leq \left(\gamma \xi^s + \delta\right) \|v_1 - v_0\| \leq r \|v_1 - v_0\|,
\end{aligned} \tag{4.2.11}$$

so (4.2.8) is true for $n = 1$. Then, we get that

$$\begin{aligned}
\|v_2 - v_0\| &\leq \|v_2 - v_1\| + \|v_1 - v_0\| \leq r \|v_1 - v_0\| + \|v_1 - v_0\| \\
&= \frac{1 - r^2}{1 - r} \xi \leq \rho,
\end{aligned} \tag{4.2.12}$$

so $v_2 \in \overline{U}(v_0, \rho)$ and v_3 exists.

Suppose that

$$\|v_{m+1} - v_m\| \leq r \|v_m - v_{m-1}\|$$

and $v_{m+1} \in \overline{U}(v_0, \rho)$ for each $m = 1, 2, \ldots, n$. Then, again as in (4.2.11) we obtain in turn that

$$\|v_{m+2} - v_{m+1}\| = \left\|A(v_{m+1})^{-1}(H(v_{m+1}) - H(v_m) - A(v_m)(v_{m+1} - v_m))\right\|$$

$$\leq \left\| A\left(v_{m+1}\right)^{-1}\left(H\left(v_{m+1}\right)-H\left(v_m\right)-H'\left(v_m\right)\left(v_{m+1}-v_m\right)\right)\right\|$$

$$+\left\| A\left(v_{m+1}\right)^{-1}\left(A\left(v_m\right)-H'\left(v_m\right)\right)\right\|$$

$$\leq \gamma \left\| v_{m+1}-v_m\right\|^{s+1}+\delta \left\| v_{m+1}-v_m\right\|$$

$$=\left(\gamma \left\| v_{m+1}-v_m\right\|^{s}+\delta\right) \left\| v_{m+1}-v_m\right\| \leq r \left\| v_{m+1}-v_m\right\| \tag{4.2.13}$$

and

$$\left\| v_{m+2}-v_0\right\| \leq \left\| v_{m+2}-v_{m+1}\right\| + \left\| v_{m+1}-v_m\right\| + \ldots + \left\| v_1-v_0\right\|$$

$$\leq \left(1+r+\ldots+r^{m+1}\right)\left\| v_1-v_0\right\| \leq \frac{1-r^{m+2}}{1-r}\left\| v_1-v_0\right\|$$

$$\leq \frac{\xi}{1-r}=\rho, \tag{4.2.14}$$

so (4.2.8) is true and $x_{m+2} \in \overline{U}\left(v_0, \rho\right)$.

Let $j \geq n$. Then, we have that

$$\left\| v_{n+j}-v_n\right\| \leq \left\| v_{n+j}-v_{n+j-1}\right\| + \ldots + \left\| v_{n+1}-v_n\right\|$$

$$\leq \left(1+r+\ldots+r^{j-1}\right)\left\| v_{n+1}-v_n\right\| = \frac{1-r^j}{1-r} r^n \left\| v_1-v_0\right\|. \tag{4.2.15}$$

Hence, by (4.2.15), $\{v_n\}$ is a complete sequence in a Banach space B_1 and as such it converges to some $v^* \in \overline{U}\left(v_0, \rho\right)$ (since $\overline{U}\left(v_0, \rho\right)$ is a closed set). By letting $j \to \infty$ in (4.2.15), we obtain (4.2.9). ■

Remark 4.2 Stronger conditions must be used to show that the limit point v^* is a solution of equation $F(x)=0$.

Theorem 4.3 *Let $H: \Omega \subseteq B_1 \to B_2$ be a Fréchet-differentiable operator. Suppose that there exist $\xi \geq 0$, $s \geq 0$, $\gamma_1 \geq 0$, $\delta_1 \geq 0$ and $\mu > 0$ such that*

$$A(x)^{-1} \in \mathcal{L}\left(B_2, B_1\right) \text{ for each } x \in \Omega_0$$

$$\left\| A(x)^{-1}\right\| \leq \mu, \quad \left\| A\left(x_0\right)^{-1} H\left(x_0\right)\right\| \leq \xi, \tag{4.2.16}$$

$$\left\| H(y)-H(x)-H'(x)(y-x)\right\| \leq \frac{\gamma_1}{\mu}\left\| y-x\right\|^{s+1} \tag{4.2.17}$$

for each $x, y \in \Omega_1$

$$\left\| A(x)-H'(x)\right\| \leq \frac{\delta_1}{\mu} \text{ for each } x \in \Omega_1, \tag{4.2.18}$$

$$\gamma_1 \xi^s + \delta_1 < 1 \tag{4.2.19}$$

and

$$\overline{U}(x_0, \rho_1) \subseteq \Omega_0, \tag{4.2.20}$$

where

$$\rho_1 = \frac{\xi}{1 - (\gamma_1 \xi^s + \delta_1)}. \tag{4.2.21}$$

Then, the conclusions of Theorem 4.1 hold for sequence $\{x_n\}$ *with* $\frac{\gamma}{\mu}, \frac{\delta}{\mu}, r_1, \rho_1$ *replacing* γ, δ, r, ρ, *respectively, where* $r_1 := \gamma_1 \xi^s + \delta_1$. *Moreover,* v^* *is a solution of the equation* $H(x) = 0$.

Proof We can write:

$$\left\| A(v_n)^{-1} \left(H(v_n) - H(v_{n-1}) - A(v_{n-1})(v_n - v_{n-1})\right) \right\| \leq$$

$$\left\| A(v_n)^{-1} \right\| \left\| H(v_n) - H(v_{n-1}) - A(v_{n-1})(v_n - v_{n-1}) \right\| \leq$$

$$\left(\gamma_1 \left\| v_n - v_{n-1} \right\|^s + \delta_1\right) \left\| v_n - v_{n-1} \right\| \leq r_1 \left\| v_n - v_{n-1} \right\|.$$

Hence, the proof of Theorem 4.1 can apply. Moreover, we obtain the estimate

$$\left\| H(v_n) \right\| = \left\| H(v_n) - H(v_{n-1}) - A(v_{n-1})(v_n - v_{n-1}) \right\| \leq$$

$$\left\| H(v_n) - H(v_{n-1}) - H'(v_{n-1})(v_n - v_{n-1}) \right\| + \left\| H(v_n) - A(v_n) \right\| \left\| v_n - v_{n-1} \right\|$$

$$\leq r_1 \left\| v_n - v_{n-1} \right\|. \tag{4.2.22}$$

By letting $n \to \infty$ in the preceding estimate we conclude that $H(v^*) = 0$. ■

Next, we present a uniqueness result for the solution v^*.

Proposition 4.4 *Suppose that the hypothesis of Theorem 4.3 and*

$$\gamma_1 \rho_1^s + \delta_1 < 1 \tag{4.2.23}$$

hold. Then, v^* *is the only solution of equation* $H(x) = 0$ *in* $\overline{U}(v_0, \rho_1)$.

Proof The existence of the solution $v^* \in \overline{U}(v_0, \rho_1)$ has been established in Theorem 4.3. Let $w^* \in \overline{U}(v_0, \rho_1)$ be such that $H(w^*) = 0$. Using (4.2.23), we can write in turn that

$$\left\| v_{n+1} - w^* \right\| = \left\| v_n - w^* - A(v_n)^{-1} H(v_n) \right\| =$$

$$\left\| A(v_n)^{-1} \left(A(v_n)\left(v_n - w^*\right) - H(v_n) + H\left(w^*\right)\right) \right\| \leq$$

$$\|A(v_n)\|^{-1} \left\| H(w^*) - H(v_n) - H'(v_n)(w^* - v_n) \right\| +$$

$$\left\| (A(v_n) - H'(v_n))(w^* - v_n) \right\| \leq$$

$$\left(\gamma_1 \left\| w^* - v_n \right\|^s + \delta_1 \right) \left\| w^* - v_n \right\| \leq$$

$$\left(\gamma_1 \rho_1^s + \delta_1 \right) \left\| v_n - w^* \right\| < \left\| v_n - w^* \right\|, \tag{4.2.24}$$

so we get that $\lim_{n \to \infty} v_n = w^*$. However, we also have that $\lim_{n \to \infty} v_n = v^*$. That is we deduce that $v^* = w^*$. ■

Remark 4.5 (i) The conditions used in this paper reduce to Newton–Mysovskii-type, if $A(x) = H'(x)$ and $s = 1$. As an example, in the case of Theorem 4.1 the sufficient semi-local convergence criterion (4.2.5) reduces to

$$\gamma \xi + \delta < 1 \tag{4.2.25}$$

used in [3–17].

(ii) Theorem 4.1 has practical value, if v^* can be shown to be a solution in some other way.

(iii) Clearly the results obtained in this paper generalize existing results for very wide class of problems by choosing operator A accordingly [6–18]. We suggest some applications to multivariate fractional calculus in the next section.

In particular, we show that condition (4.2.18) holds, if the right hand side of (4.3.9) (or (4.3.18)) that follow are replaced by $\frac{\delta_1}{\mu}$. Other applications and choices for A can be found in [6–18].

4.3 Applications to Multivariate Fractional Calculus

(I) Let $\alpha = (\alpha_1, ..., \alpha_k)$, $\alpha_i \in (0, 1)$, $i = 1, ..., k \in \mathbb{N}$, and $G \in C^{k-1}\left(\prod_{i=1}^k [c_i, d_i]\right)$, such that

$$0 \neq \frac{\partial^k G}{\partial x_1 ... \partial x_k} \in C\left(\prod_{i=1}^k [c_i, d_i]\right).$$

Here we consider the multivariate left Caputo type fractional mixed partial derivative of order α :

$$D_{*a}^{\alpha} G(x) = \frac{1}{\prod_{i=1}^k \Gamma(1 - \alpha_i)} \cdot$$

$$\int_{a_1}^{x_1} ... \int_{a_k}^{x_k} \prod_{i=1}^k (x_i - t_i)^{-\alpha_i} \frac{\partial^k G(t_1, ..., t_k)}{\partial t_1 ... \partial t_k} dt_1 ... dt_k, \tag{4.3.1}$$

where Γ is the gamma function, $a = (a_1, ..., a_k)$, $\forall\, x = (x_1, ..., x_k) \in \prod_{i=1}^{k} [a_i, d_i]$. Here $c_i < a_i < b_i < d_i$, $i = 1, ..., k$.

If any $x_i \in [c_i, a_i]$, $i \in \{1, ..., k\}$, we define $D_{*a}^{\alpha} G(x) = 0$.

Next we consider $a_i < a_i^* < b_i$, and $x_i \in [a_i^*, b_i]$, also $x_{0i} \in (c_i, a_i)$, for all $i = 1, ..., k$. That is $x \in \prod_{i=1}^{k} [a_i^*, b_i]$, and $x_0 = (x_{01}, ..., x_{0k}) \in \prod_{i=1}^{k} (c_i, a_i)$.

We define the function

$$A_1(x) := \prod_{i=1}^{k} \left(\frac{\Gamma(2-\alpha_i)}{(x_i - a_i)^{1-\alpha_i}} \right) D_{*a}^{\alpha} G(x), \quad \forall x \in \prod_{i=1}^{k} [a_i^*, b_i]. \tag{4.3.2}$$

Notice that $A_1(a)$ is undefined. We see that

$$\left| A_1(x) - \frac{\partial^k G(x_1, ..., x_k)}{\partial x_1 ... \partial x_k} \right| =$$

$$\left| \prod_{i=1}^{k} \left(\frac{\Gamma(2-\alpha_i)}{(x_i - a_i)^{1-\alpha_i}} \right) D_{*a}^{\alpha} G(x) - \frac{\partial^k G(x_1, ..., x_k)}{\partial x_1 ... \partial x_k} \right| = \tag{4.3.3}$$

$$\left| \prod_{i=1}^{k} \left(\frac{\Gamma(2-\alpha_i)}{(x_i - a_i)^{1-\alpha_i}} \right) \frac{1}{\prod_{i=1}^{k} \Gamma(1-\alpha_i)} \int_{a_1}^{x_1} ... \int_{a_k}^{x_k} \prod_{i=1}^{k} (x_i - t_i)^{-\alpha_i} \cdot \right.$$

$$\left. \frac{\partial^k G(t_1, ..., t_k)}{\partial t_1 ... \partial t_k} dt_1 ... dt_k - \frac{\partial^k G(x_1, ..., x_k)}{\partial x_1 ... \partial x_k} \right| =$$

$$\left| \prod_{i=1}^{k} \frac{\Gamma(2-\alpha_i)}{(x_i - a_i)^{1-\alpha_i} \Gamma(1-\alpha_i)} \int_{a_1}^{x_1} ... \int_{a_k}^{x_k} \prod_{i=1}^{k} (x_i - t_i)^{-\alpha_i} \frac{\partial^k G(t_1, ..., t_k)}{\partial t_1 ... \partial t_k} dt_1 ... dt_k \right.$$

$$\left. - \prod_{i=1}^{k} \frac{\Gamma(2-\alpha_i)}{(x_i - a_i)^{1-\alpha_i}} \frac{(x_i - a_i)^{1-\alpha_i}}{\Gamma(2-\alpha_i)} \frac{\partial^k G(x_1, ..., x_k)}{\partial x_1 ... \partial x_k} \right| = \tag{4.3.4}$$

$$\prod_{i=1}^{k} \frac{\Gamma(2-\alpha_i)}{(x_i - a_i)^{1-\alpha_i}} \left| \frac{1}{\prod_{i=1}^{k} \Gamma(1-\alpha_i)} \int_{a_1}^{x_1} ... \int_{a_k}^{x_k} \prod_{i=1}^{k} (x_i - t_i)^{-\alpha_i} \cdot \right.$$

$$\frac{\partial^k G(t_1, ..., t_k)}{\partial t_1 ... \partial t_k} dt_1 ... dt_k - \frac{1}{\prod_{i=1}^{k} \Gamma(1-\alpha_i)} \int_{a_1}^{x_1} ... \int_{a_k}^{x_k} \prod_{i=1}^{k} (x_i - t_i)^{-\alpha_i} \cdot$$

$$\left. \frac{\partial^k G(x_1, ..., x_k)}{\partial x_1 ... \partial x_k} dt_1 ... dt_k \right| =$$

$$\prod_{i=1}^{k}\frac{(1-\alpha_i)}{(x_i-a_i)^{1-\alpha_i}}\left|\int_{a_1}^{x_1}\dots\int_{a_k}^{x_k}\prod_{i=1}^{k}(x_i-t_i)^{-\alpha_i}\cdot\right.$$

$$\left.\left(\frac{\partial^k G(t_1,\dots,t_k)}{\partial t_1\dots\partial t_k}-\frac{\partial^k G(x_1,\dots,x_k)}{\partial x_1\dots\partial x_k}\right)dt_1\dots dt_k\right|\leq \tag{4.3.5}$$

$$\prod_{i=1}^{k}\frac{(1-\alpha_i)}{(x_i-a_i)^{1-\alpha_i}}\int_{a_1}^{x_1}\dots\int_{a_k}^{x_k}\prod_{i=1}^{k}(x_i-t_i)^{-\alpha_i}\cdot$$

$$\left|\frac{\partial^k G(t_1,\dots,t_k)}{\partial t_1\dots\partial t_k}-\frac{\partial^k G(x_1,\dots,x_k)}{\partial x_1\dots\partial x_k}\right|dt_1\dots dt_k$$

(we assume that there exists $K>0$) :

$$\left|\frac{\partial^k G(t_1,\dots,t_k)}{\partial t_1\dots\partial t_k}-\frac{\partial^k G(x_1,\dots,x_k)}{\partial x_1\dots\partial x_k}\right|\leq K\left(\prod_{i=1}^{k}|t_i-x_i|\right), \tag{4.3.6}$$

$\forall\, t=(t_1,\dots,t_k), x=(x_1,\dots,x_k)\in\prod_{i=1}^{k}[c_i,d_i])$

$$\leq K\prod_{i=1}^{k}\frac{(1-\alpha_i)}{(x_i-a_i)^{1-\alpha_i}}\int_{a_1}^{x_1}\dots\int_{a_k}^{x_k}\prod_{i=1}^{k}(x_i-t_i)^{-\alpha_i}\prod_{i=1}^{k}(x_i-t_i)\,dt_1\dots dt_k= \tag{4.3.7}$$

$$K\prod_{i=1}^{k}\frac{(1-\alpha_i)}{(x_i-a_i)^{1-\alpha_i}}\int_{a_1}^{x_1}\dots\int_{a_k}^{x_k}\prod_{i=1}^{k}(x_i-t_i)^{1-\alpha_i}\,dt_1\dots dt_k=$$

$$K\prod_{i=1}^{k}\frac{(1-\alpha_i)}{(x_i-a_i)^{1-\alpha_i}}\prod_{i=1}^{k}\frac{(x_i-a_i)^{2-\alpha_i}}{(2-\alpha_i)}=$$

$$K\prod_{i=1}^{k}\left(\frac{1-\alpha_i}{2-\alpha_i}\right)(x_i-a_i).$$

We have proved that

$$\left|A_1(x)-\frac{\partial^k G(x_1,\dots,x_k)}{\partial x_1\dots\partial x_k}\right|\leq$$

$$\left(\prod_{i=1}^{k}\left(\frac{1-\alpha_i}{2-\alpha_i}\right)\right)K\left(\prod_{i=1}^{k}(x_i-a_i)\right)\leq \tag{4.3.8}$$

$$\left(\prod_{i=1}^{k}\left(\frac{1-\alpha_i}{2-\alpha_i}\right)\right)K\left(\prod_{i=1}^{k}(b_i-a_i)\right),$$

$\forall\, x \in \prod_{i=1}^{k}\left[a_i^*, b_i\right]$.

In particular, it holds that

$$\left|A_1(x)-\frac{\partial^k G(x_1,...,x_k)}{\partial x_1...\partial x_k}\right| \le$$

$$\left(\prod_{i=1}^{k}\left(\frac{1-\alpha_i}{2-\alpha_i}\right)\right)K\left(\prod_{i=1}^{k}(x_i-x_{0i})\right), \tag{4.3.9}$$

where $x_0=(x_{01},...,x_{0k})\in\prod_{i=1}^{k}(c_i,a_i)$, $\forall\, x=(x_1,...,x_k)\in\prod_{i=1}^{k}\left[a_i^*,b_i\right]$.

(II) Let $\alpha=(\alpha_1,...,\alpha_k)$, $\alpha_i\in(0,1)$, $i=1,...,k\in\mathbb{N}$, and $G\in C^{k-1}\left(\prod_{i=1}^{k}[c_i,d_i]\right)$, such that

$$0\neq\frac{\partial^k G}{\partial x_1...\partial x_k}\in C\left(\prod_{i=1}^{k}[c_i,d_i]\right).$$

Here we consider the multivariate right Caputo type fractional mixed partial derivative of order α :

$$D_{b-}^{\alpha}G(x)=\frac{(-1)^k}{\prod_{i=1}^{k}\Gamma(1-\alpha_i)}\cdot$$

$$\int_{x_1}^{b_1}...\int_{x_k}^{b_k}\prod_{i=1}^{k}(t_i-x_i)^{-\alpha_i}\frac{\partial^k G(t_1,...,t_k)}{\partial t_1...\partial t_k}dt_1...dt_k, \tag{4.3.10}$$

where Γ is the gamma function, $b=(b_1,...,b_k)$, $\forall\, x=(x_1,...,x_k)\in\prod_{i=1}^{k}[c_i,b_i]$. Here $c_i<a_i<b_i<d_i$, $i=1,...,k$.

If any $x_i\in[b_i,d_i]$, $i\in\{1,...,k\}$, we define $D_{b-}^{\alpha}G(x)=0$.

Next we consider $a_i<b_i^*<b_i$, and $x_i\in\left[a_i,b_i^*\right]$, also $x_{0i}\in(b_i,d_i)$, for all $i=1,...,k$. That is $x\in\prod_{i=1}^{k}\left[a_i,b_i^*\right]$, and $x_0=(x_{01},\ldots,x_{0k})\in\prod_{i=1}^{k}(b_i,d_i)$.

We define the function

$$A_2(x):=(-1)^k\left(\prod_{i=1}^{k}\left(\frac{\Gamma(2-\alpha_i)}{(b_i-x_i)^{1-\alpha_i}}\right)\right)D_{b-}^{\alpha}G(x),\ \ \forall\, x\in\prod_{i=1}^{k}\left[a_i,b_i^*\right]. \tag{4.3.11}$$

Notice that $A_2(b)$ is undefined. We see that

$$\left|A_2(x)-\frac{\partial^k G(x_1,...,x_k)}{\partial x_1...\partial x_k}\right| =$$

$$\left|(-1)^k \prod_{i=1}^{k}\left(\frac{\Gamma(2-\alpha_i)}{(b_i-x_i)^{1-\alpha_i}}\right) D_{b-}^{\alpha} G(x) - \frac{\partial^k G(x_1,...,x_k)}{\partial x_1 ... \partial x_k}\right| = \tag{4.3.12}$$

$$\left|(-1)^k \prod_{i=1}^{k}\left(\frac{\Gamma(2-\alpha_i)}{(b_i-x_i)^{1-\alpha_i}}\right) \frac{(-1)^k}{\prod_{i=1}^{k}\Gamma(1-\alpha_i)} \int_{x_1}^{b_1} ... \int_{x_k}^{b_k} \prod_{i=1}^{k} (t_i-x_i)^{-\alpha_i} \cdot \right.$$

$$\left.\frac{\partial^k G(t_1,...,t_k)}{\partial t_1 ... \partial t_k} dt_1 ... dt_k - \frac{\partial^k G(x_1,...,x_k)}{\partial x_1 ... \partial x_k}\right| =$$

$$\left|\prod_{i=1}^{k} \frac{\Gamma(2-\alpha_i)}{(b_i-x_i)^{1-\alpha_i}\,\Gamma(1-\alpha_i)} \int_{x_1}^{b_1} ... \int_{x_k}^{b_k} \prod_{i=1}^{k} (t_i-x_i)^{-\alpha_i} \frac{\partial^k G(t_1,...,t_k)}{\partial t_1 ... \partial t_k} dt_1 ... dt_k\right.$$

$$\left.-\prod_{i=1}^{k} \frac{\Gamma(2-\alpha_i)}{(b_i-x_i)^{1-\alpha_i}} \frac{(b_i-x_i)^{1-\alpha_i}}{\Gamma(2-\alpha_i)} \frac{\partial^k G(x_1,...,x_k)}{\partial x_1 ... \partial x_k}\right| = \tag{4.3.13}$$

$$\prod_{i=1}^{k} \frac{\Gamma(2-\alpha_i)}{(b_i-x_i)^{1-\alpha_i}} \left|\frac{1}{\prod_{i=1}^{k}\Gamma(1-\alpha_i)} \int_{x_1}^{b_1} ... \int_{x_k}^{b_k} \prod_{i=1}^{k} (t_i-x_i)^{-\alpha_i} \cdot \right.$$

$$\frac{\partial^k G(t_1,...,t_k)}{\partial t_1 ... \partial t_k} dt_1 ... dt_k - \frac{1}{\prod_{i=1}^{k}\Gamma(1-\alpha_i)} \int_{x_1}^{b_1} ... \int_{x_k}^{b_k} \prod_{i=1}^{k} (t_i-x_i)^{-\alpha_i} \cdot$$

$$\left.\frac{\partial^k G(x_1,...,x_k)}{\partial x_1 ... \partial x_k} dt_1 ... dt_k\right| =$$

$$\prod_{i=1}^{k} \frac{(1-\alpha_i)}{(b_i-x_i)^{1-\alpha_i}} \left|\int_{x_1}^{b_1} ... \int_{x_k}^{b_k} \prod_{i=1}^{k} (t_i-x_i)^{-\alpha_i} \cdot \right. \tag{4.3.14}$$

$$\left.\left(\frac{\partial^k G(t_1,...,t_k)}{\partial t_1 ... \partial t_k} - \frac{\partial^k G(x_1,...,x_k)}{\partial x_1 ... \partial x_k}\right) dt_1 ... dt_k\right| \leq$$

$$\prod_{i=1}^{k} \frac{(1-\alpha_i)}{(b_i-x_i)^{1-\alpha_i}} \int_{x_1}^{b_1} ... \int_{x_k}^{b_k} \prod_{i=1}^{k} (t_i-x_i)^{-\alpha_i} \cdot$$

$$\left|\frac{\partial^k G(t_1,...,t_k)}{\partial t_1 ... \partial t_k} - \frac{\partial^k G(x_1,...,x_k)}{\partial x_1 ... \partial x_k}\right| dt_1 ... dt_k$$

(we assume that there exists $K > 0$:

$$\left| \frac{\partial^k G(t_1, ..., t_k)}{\partial t_1 ... \partial t_k} - \frac{\partial^k G(x_1, ..., x_k)}{\partial x_1 ... \partial x_k} \right| \leq K \left(\prod_{i=1}^{k} |t_i - x_i| \right), \tag{4.3.15}$$

$\forall\, t = (t_1, ..., t_k), x = (x_1, ..., x_k) \in \prod_{i=1}^{k} [c_i, d_i])$

$$\leq K \prod_{i=1}^{k} \frac{(1-\alpha_i)}{(b_i - x_i)^{1-\alpha_i}} \int_{x_1}^{b_1} ... \int_{x_k}^{b_k} \prod_{i=1}^{k} (t_i - x_i)^{-\alpha_i} \prod_{i=1}^{k} (t_i - x_i)\, dt_1 ... dt_k =$$

$$K \prod_{i=1}^{k} \frac{(1-\alpha_i)}{(b_i - x_i)^{1-\alpha_i}} \int_{x_1}^{b_1} ... \int_{x_k}^{b_k} \prod_{i=1}^{k} (t_i - x_i)^{1-\alpha_i}\, dt_1 ... dt_k = \tag{4.3.16}$$

$$K \prod_{i=1}^{k} \frac{(1-\alpha_i)}{(b_i - x_i)^{1-\alpha_i}} \prod_{i=1}^{k} \frac{(b_i - x_i)^{2-\alpha_i}}{(2-\alpha_i)} =$$

$$K \prod_{i=1}^{k} \left(\frac{1-\alpha_i}{2-\alpha_i} \right) (b_i - x_i).$$

We have proved that

$$\left| A_2(x) - \frac{\partial^k G(x_1, ..., x_k)}{\partial x_1 ... \partial x_k} \right| \leq$$

$$\left(\prod_{i=1}^{k} \left(\frac{1-\alpha_i}{2-\alpha_i} \right) \right) K \left(\prod_{i=1}^{k} (b_i - x_i) \right) \leq \tag{4.3.17}$$

$$\left(\prod_{i=1}^{k} \left(\frac{1-\alpha_i}{2-\alpha_i} \right) \right) K \left(\prod_{i=1}^{k} (b_i - a_i) \right),$$

$\forall\, x \in \prod_{i=1}^{k} \left[a_i, b_i^* \right]$.

In particular, it holds that

$$\left| A_2(x) - \frac{\partial^k G(x_1, ..., x_k)}{\partial x_1 ... \partial x_k} \right| \leq$$

$$\left(\prod_{i=1}^{k} \left(\frac{1-\alpha_i}{2-\alpha_i} \right) \right) K \left(\prod_{i=1}^{k} (x_{0i} - x_i) \right), \tag{4.3.18}$$

where $x_0 = (x_{01}, ..., x_{0k}) \in \prod_{i=1}^{k} (b_i, d_i)$, $\forall\, x = (x_1, ..., x_k) \in \prod_{i=1}^{k} \left[a_i, b_i^* \right]$.

References

1. S. Amat, S. Busquier, S. Plaza, Chaotic dynamics of a third-order Newton-type method. J. Math. Anal. Appl. **366**(1), 164–174 (2010)
2. G. Anastassiou, *Fractional Differentiation Inequalities* (Springer, New York, 2009)
3. G. Anastassiou, Fractional representation formulae and right fractional inequalities. Math. Comput. Model. **54**(10–12), 3098–3115 (2011)
4. G. Anastassiou, *Intelligent Mathematics: Computational Analysis* (Springer, Heidelberg, 2011)
5. G. Anastassiou, *Advanced Inequalities* (World Scientific Publ. Corp, Singapore, 2011)
6. G. Anastassiou, I.K. Argyros, *Intelligent Numerical Methods: Applications to Fractional Calculus*, Studies in Computational Intelligence (Springer, Heidelberg, 2016)
7. G. Anastassiou, I. Argyros, Newton-like Methods and their Applications in Multivariate Fractional Calculus, submitted for publication (2015)
8. I.K. Argyros, A unifying local-semilocal convergence analysis and applications for two-point Newton-like methods in Banach space. J. Math. Anal. Appl. **298**, 374–397 (2004)
9. I.K. Argyros, *Convergence and Applications of Newton-Type Iterations* (Springer, New York, 2008)
10. I.K. Argyros, On a class of Newton-like methods for solving nonlinear equations. J. Comput. Appl. Math. **228**, 115–122 (2009)
11. I.K. Argyros, A semilocal convergence analysis for directional Newton methods. AMS J. **80**, 327–343 (2011)
12. I.K. Argyros, Y.J. Cho, S. Hilout, *Numerical Methods for Equations and its Applications* (CRC Press/Taylor and Fracncis, New York, 2012)
13. I.K. Argyros, S. Hilout, Weaker conditions for the convergence of Newton's method. J. Complex. **28**, 364–387 (2012)
14. J.A. Ezquérro, J.M. Gutiérrez, M.A. Hernández, N. Romero, M.J. Rubio, The Newton method: from Newton to Kantorovich (Spanish). Gac. R. Soc. Mat. Esp. **13**, 53–76 (2010)
15. J.A. Ezquérro, M.A. Hernández, Newton-type methods of high order and domains of semilocal and global convergence. Appl. Math. Comput. **214**(1), 142–154 (2009)
16. L.V. Kantorovich, G.P. Akilov, *Functional Analysis in Normed Spaces* (Pergamon Press, New York, 1964)
17. A.A. Magréñán, Different anomalies in a Jarratt family of iterative root finding methods. Appl. Math. Comput. **233**, 29–38 (2014)
18. A.A. Magréñán, A new tool to study real dynamics: the convergence plane. Appl. Math. Comput. **248**, 215–224 (2014)

Chapter 5
Implicit Iterative Algorithms and Their Applications in Multivariate Calculus

We present a semilocal convergence analysis of implicit iterative methods for approximating a solution of an equation in a Banach space setting. The sufficient conditions are more general even in the explicit convergence case leading to a larger convergence domain, tighter error estimates on the distances involved and an at least as precise information on the location of the solution than in earlier studies. Applications are suggested in several areas including multivariate calculus. It follows [7].

5.1 Introduction

Numerous problems can be formulated as an equation of the form

$$G(x) = 0 \tag{5.1.1}$$

using Mathematical Modelling [1–6, 11, 12, 15–20], where G is a continuous operator defined on a subset D of a Banach space X with values in a Banach space Y.

The solutions x^* of Eq. (5.1.1) can rarely be found in closed form. That explains why most solution methods for these equations are usually iterative.

In the present chapter we study the semi-local convergence analysis of iterative method defined implicitly for $x_0 \in D$ by

$$G(x_n) + A(x_{n+1} - x_n) = 0, \tag{5.1.2}$$

where $A : D \to Y$ is a given operator with some properties (to be precised in Sect. 5.2) such that sequence $\{x_n\}$ generated implicitly by method (5.1.2) converges to x^*. Clearly method (5.1.2) is very general and includes the Newton, Steffensen, Secant, Newton-like, secant-like and other popular explicit methods for special choices of operator A [6–20].

G.A. Anastassiou and I.K. Argyros, *Intelligent Numerical Methods II: Applications to Multivariate Fractional Calculus*, Studies in Computational Intelligence 649, DOI 10.1007/978-3-319-33606-0_5

As far as we know results in the literature deal with the special case when $A := A(\cdot)$ is a linear operator, e.g. such as $A(\cdot) : D \to L(X, Y)$ the space of bounded linear operators from X into Y. In those cases one hopes that the implicit method (5.1.2) can be written in an explicit form

$$x_{n+1} = x_n - A_n^{-1} G(x_n), \quad A_n = A(x_n). \tag{5.1.3}$$

We refer the reader to [1, 6–20] and the references there in for results concerning the local as well as the semi-local convergence of method (5.1.3).

The rest of the chapter is organized as follows. Section 5.2 contains the semi-local convergence of some implicit and explicit method. Applications are suggested at the end of Sect. 5.2 as well as in Sect. 5.3 where in particular cases involving multivariate calculus are provided.

In the end of this paper $U(x, \rho)$, $\overline{U}(x, \rho)$ stand, respectively for the open and closed balls in X, respectively with center $x \in D$ and of radius $\rho > 0$.

5.2 Semilocal Convergence

We present the semilocal convergence analysis of the method (5.1.2) in this section starting with a very general result.

Theorem 5.1 *Let $G : D \to Y$ be a continuous operator and $A : D \to Y$ be an operator. Suppose:*

there exist $\gamma > 0$ and $\delta \geq 1$ such that for each $x, y \in D$

$$\|G(y) - G(x) - A(y - x)\| \leq \gamma \|y - x\|^{\delta}; \tag{5.2.1}$$

sequence $\{x_n\}$ generated for $x_0 \in D$ by method (5.1.2) is well defined in D, remains in D for each $n = 0, 1, 2, \ldots$ and is complete in D.

Then, sequence $\{x_n\}$ converges to a solution $x^* \in \overline{D}$ of equation $G(x) = 0$.

Proof In view of the definition of the sequence $\{x_k\}$ and condition (5.2.1), we get from the identity

$$G(x_{k+1}) = G(x_{k+1}) - G(x_k) - A(x_{k+1} - x_k) \tag{5.2.2}$$

that

$$\|G(x_{k+1})\| = \|G(x_{k+1}) - G(x_k) - A(x_{k+1} - x_k)\| \leq \gamma \|x_{k+1} - x_k\|^{\delta}. \tag{5.2.3}$$

By hypothesis $\{x_k\}$ is a complete sequence in a Banach space X and as such it converges to some $x^* \in \overline{D}$ (since $\overline{D}$ is a closed set). Using the continuity of G and by letting $k \to \infty$ in (5.2.3), we deduce that $G(x^*) = 0$. ∎

Remark 5.2 Theorem 5.1 is very usefull in cases the existence and completeness of sequence $\{x_n\}$ has been determined in some way. Results of this type have been given in [1, 6–20] and the references there in, when A is a linear operator. That is $A := A(\cdot) : D \to L(X, Z)$, where $Z = X$ or $Z = Y$.

Let $L \in L(Y, X)$ and $A : X \to X$. Set $F = LG$. The iterates $\{x_n\}$ will be determined through a fixed point problem as follows:

$$x_{n+1} = x_n + y_n, \quad A(y_n) + F(x_n) = 0 \Leftrightarrow y_n = Q(y_n) := y_n - A(y_n) - F(x_n). \tag{5.2.4}$$

Let $R > 0$. Define $R_0 := \sup\{t \in [0, R] : U(x_0, t) \subseteq D\}$ and $D_0 = \overline{U}(x_0, R_0)$.

Approximate solutions of the equation $F(x) = 0$ using iterative scheme (5.2.4) have been given in [6].

Remark 5.3 If $A = I$, the identity operator $I : X \to X$, then (5.2.4) reduces to Picard's iteration $z_{n+1} = F(z_n)$, $z_0 = x_0$ [16] for computing fixed points of operator F. Many other choices for A are also possible [6–20].

So far the sequence $\{x_n\}$ was computed implicitly. Next, sequence $\{x_n\}$ is defined explicitly by

$$x_{n+1} = x_n - B_n^{-1} G(x_n), \tag{5.2.5}$$

where $x_0 \in D$ is an initial point, $B_n := B(x_n)$, and $B_n^{-1} \in L(Y, X)$ for each $n = 0, 1, 2, \ldots$.

We state the following semilocal convergence result for method (5.2.5).

Theorem 5.4 *Let $G : D_0 \to Y$ be a continuous operator and let $\{B_n\} \in L(X, Y)$ be a sequence such that*

$$\left\| B(x)^{-1} B_0 \right\| \leq c, \tag{5.2.6}$$

$\{B_n^{-1}\} \in L(Y, X)$ for each $x \in \overline{U}(x_0, R_1)$, and some $R_1, c > 0$.

Moreover, suppose: there exist $d \geq 0$ and a function $q : [0, R_1) \to [0, +\infty)$ continuous and non-decreasing such that

$$\left\| B_0^{-1} G(x_0) \right\| \leq d; \tag{5.2.7}$$

$$\left\| B_0^{-1} (G(y) - G(x) - B(x)(y - x)) \right\| \leq q(\|x - x_0\|) \|y - x\| \tag{5.2.8}$$

for each $x, y \in D_1 := D_0 \cap \overline{U}(x_0, R_1)$.

Let $R_2 = \min\{R_0, R_1\}$.

$$cq(t) < 1 \quad \text{for each } t \in [0, R_2). \tag{5.2.9}$$

Let also

$$\mu := \sup\{t \in [0, R_2] : cq(t) < 1\} \tag{5.2.10}$$

and

$$R_3 := \frac{d}{1-\mu} \leq R_2. \tag{5.2.11}$$

Then, the sequence $\{x_n\}$ *generated for* $x_0 \in D$ *by method (5.2.5) is well defined in* $\overline{U}(x_0, R_3)$, *remains in* $\overline{U}(x_0, R_3)$ *for each* $n = 0, 1, 2, ...$ *and converges to a solution* $x^* \in \overline{U}(x_0, R_3)$ *of equation* $G(x) = 0$.

Proof Notice that $\mu \in [0, 1)$ by (5.2.9) and (5.2.10).

Point x_1 is well defined by method (5.2.5) for $n = 0$ and since $B_0^{-1} \in L(Y, X)$. In view of (5.2.7) and (5.2.11):

$$\|x_1 - x_0\| = d < R_3, \tag{5.2.12}$$

so $x_1 \in U(x_0, R_3) \subseteq \overline{U}(x_0, R_3)$, $B(x_1)^{-1} \in L(Y, X)$ and x_2 is well defined by method (5.2.5) for $n = 1$. We can write by method (5.2.5) that

$$G(x_1) = G(x_1) - G(x_0) - B(x_1)(x_2 - x_1). \tag{5.2.13}$$

Then, by (5.2.5), (5.2.7)–(5.2.10), we obtain in turn that

$$\begin{aligned} \|x_2 - x_1\| &= \left\|\left(B(x_1)^{-1} B_0\right)\left(B_0^{-1} G(x_1)\right)\right\| \leq \left\|B(x_1)^{-1} B_0\right\| \left\|B_0^{-1} G(x_1)\right\| \\ &\leq cq(\|x_1 - x_0\|)\|x_1 - x_0\| \leq \mu\|x_1 - x_0\| \leq \mu d \end{aligned} \tag{5.2.14}$$

and

$$\|x_2 - x_0\| \leq \|x_2 - x_1\| + \|x_1 - x_0\| \leq \mu d + d = \frac{1-\mu^2}{1-\mu} d < \frac{d}{1-\mu} = R_3, \tag{5.2.15}$$

so $x_2 \in U(x_0, R_3) \subseteq \overline{U}(x_0, R_3)$, $B(x_2)^{-1} \in L(Y, X)$ and x_3 is well defined by method (5.2.5) for $n = 2$. Continuing in an analogous way and using induction, we get that

$$\|x_{n+1} - x_n\| \leq \mu\|x_n - x_{n-1}\| \leq \mu^2\|x_{n-1} - x_{n-2}\| \leq \mu^n\|x_1 - x_0\| \leq \mu^n d \tag{5.2.16}$$

and

$$\|x_{n+1} - x_0\| \leq \sum_{i=0}^{n} \|x_{i+1} - x_i\| \leq \left(1 + \mu + ... + \mu^n\right) d = \tag{5.2.17}$$

$$\frac{1-\mu^{n+1}}{1-\mu}d < \frac{d}{1-\mu} = R_3,$$

so $x_{n+1} \in U(x_0, R_3)$, $B(x_{n+1})^{-1} \in L(Y, X)$ and x_{n+2} is well defined.

We also have that

$$\|x_{n+m} - x_n\| \leq \|x_{n+m} - x_{n+m-1}\| + \|x_{n+m-1} - x_{n+m-2}\| + \ldots + \|x_{n+1} - x_n\|$$

$$\leq \left(\mu^{n+m} + \ldots + \mu^n\right) d = \frac{1-\mu^m}{1-\mu}\mu^n d. \tag{5.2.18}$$

Estimate (5.2.18) shows the sequence $\{x_n\}$ is complete in a Banach space X and as such it converges to some $x^* \in \overline{U}(x_0, R_3)$ (since $\overline{U}(x_0, R_3)$ is a closed set). By letting $n \to \infty$ in the estimate

$$\|G(x_{n+1})\| \leq q(R_2)\|x_{n+1} - x_n\| \tag{5.2.19}$$

and using the continuity of G, we conclude that $G(x^*) = 0$ and $x^* \in \overline{U}(x_0, R_3)$. ■

Remark 5.5 Earlier results concerning the convergence of method (5.2.5) were using condition (5.2.8) as

$$\left\| B_0^{-1}\left(G(y) - G(x) - B(x)(y - x)\right)\right\| \leq q_1(\|x - x_0\|)\|y - x\| \tag{5.2.20}$$

for each $x, y \in D_0$.

However, $D_1 \subseteq D_0$ holds. Therefore,

$$q(t) \leq q_1(t) \quad \text{for each } t \in [0, R_1] \tag{5.2.21}$$

leading to tighter sufficient semilocal convergence conditions, tighter error estimates on the distances and an at least as precise information on the location of the solution x^*, since

$$cq_1(t) < 1 \Rightarrow cq(t) < 1 \tag{5.2.22}$$

and

$$\mu \leq \mu_1 := \sup\{t \in [0, R_2] : cq_1(t) < 1\}, \tag{5.2.23}$$

but not necessarily vice versa, unless if $D_0 = D_1$.

In the next section, we suggest some applications in multivariate calculus concerning Theorem 5.1. Applications of the other results can also be found in [1, 6–20].

5.3 Application to Multivariate Calculus

Let $f \in C^1\left(\prod_{j=1}^{k}\left[a_j, b_j\right]\right)$, $k \geq 2$; $x = (x_1, \ldots, x_k)$, $y = (y_1, \ldots, y_k) \in \prod_{j=1}^{k}\left[a_j, b_j\right]$. By [5], pp. 284–286, we obtain

$$f(x) - f(y) = \sum_{i=1}^{k} (x_i - y_i) \int_0^1 \frac{\partial f}{\partial x_i}(y + \theta(x - y))\, d\theta. \tag{5.3.1}$$

We assume that

$$\left| \frac{\partial f}{\partial x_i}(x) - \frac{\partial f}{\partial x_i}(y) \right| \leq \lambda \|x - y\|_{l_1}, \tag{5.3.2}$$

where $0 < \lambda < 1$, $\forall\, x, y \in \prod_{j=1}^{k}\left[a_j, b_j\right]$, for all $i = 1, \ldots, k$.

Here, we take as

$$A_0(f)(x) = \left(\frac{\partial f}{\partial x_1}(x), \ldots, \frac{\partial f}{\partial x_k}(x) \right), \tag{5.3.3}$$

all $x \in \prod_{j=1}^{k}\left[a_j, b_j\right]$.

Notice that $A_0 : C^1\left(\prod_{j=1}^{k}\left[a_j, b_j\right]\right) \to \left(C\left(\prod_{j=1}^{k}\left[a_j, b_j\right]\right)\right)^k$ is a linear operator.

Therefore it holds (below "$\cdot$" denotes the inner product)

$$|f(x) - f(y) - (A_0(f))(x) \cdot (x - y)| \overset{\text{(by (5.3.1),(5.3.3))}}{=} \tag{5.3.4}$$

$$\left| \sum_{i=1}^{k} (x_i - y_i) \int_0^1 \frac{\partial f}{\partial x_i}(y + \theta(x - y))\, d\theta - \sum_{i=1}^{k} (x_i - y_i) \frac{\partial f}{\partial x_i}(x) \right| =$$

$$\left| \sum_{i=1}^{k} (x_i - y_i) \int_0^1 \frac{\partial f}{\partial x_i}(y + \theta(x - y))\, d\theta - \sum_{i=1}^{k} (x_i - y_i) \left(\int_0^1 \frac{\partial f(x)}{\partial x_i} d\theta \right) \right| =$$

$$\left| \sum_{i=1}^{k} (x_i - y_i) \int_0^1 \left(\frac{\partial f}{\partial x_i}(y + \theta(x - y)) - \frac{\partial f(x)}{\partial x_i} \right) d\theta \right| \leq \tag{5.3.5}$$

$$\sum_{i=1}^{k} |x_i - y_i| \int_0^1 \left| \frac{\partial f}{\partial x_i}(y + \theta(x - y)) - \frac{\partial f(x)}{\partial x_i} \right| d\theta \overset{(5.3.2)}{\leq}$$

$$\lambda\left(\sum_{i=1}^{k}|x_i - y_i|\right)\int_0^1 \|y + \theta(x-y) - x\|_{l_1}\, d\theta = \tag{5.3.6}$$

$$\lambda\|x-y\|_{l_1}\int_0^1 (1-\theta)\|x-y\|_{l_1}\, d\theta = \lambda\|x-y\|_{l_1}^2\int_0^1 (1-\theta)\, d\theta =$$

$$\lambda\|x-y\|_{l_1}^2\left(\left.\frac{(1-\theta)^2}{2}\right|_1^0\right) = \frac{\lambda\|x-y\|_{l_1}^2}{2}. \tag{5.3.7}$$

We have proved that

$$|f(x) - f(y) - (A_0(f))(x)\cdot(x-y)| \leq \frac{\lambda\|x-y\|_{l_1}^2}{2}, \quad \forall x, y \in \prod_{j=1}^{k}[a_j, b_j], \tag{5.3.8}$$

a condition needed to solve numerically $f(x) = 0$. In particular, according to condition (5.2.1), we must choose $\gamma = \frac{\lambda}{2}$ and $\delta = 2$.

References

1. S. Amat, S. Busquier, S. Plaza, Chaotic dynamics of a third-order Newton-type method. J. Math. Anal. Appl. **366**(1), 164–174 (2010)
2. G. Anastassiou, *Fractional Differentiation Inequalities* (Springer, New York, 2009)
3. G. Anastassiou, Fractional representation formulae and right fractional inequalities. Mathematical and Computer Modelling **54**(10–12), 3098–3115 (2011)
4. G. Anastassiou, *Intelligent Mathematics: Computational Analysis* (Springer, Heidelberg, 2011)
5. G. Anastassiou, *Advanced Inequalities* (World Scientific Publisher Company, Singapore, 2011)
6. G. Anastassiou, I.K. Argyros, *Studies in Computational Intelligence. Intelligent Numerical Methods: Applications to Fractional Calculus*, vol. 624 (Springer, Heidelberg, 2016)
7. G. Anastassiou, I. Argyros, Implicit iterative methods for solving equations with applications in multivariate calculus, submitted for publication (2015)
8. I.K. Argyros, A unifying local-semilocal convergence analysis and applications for two-point Newton-like methods in Banach space. J. Math. Anal. Appl. **298**, 374–397 (2004)
9. I.K. Argyros, *Convergence and Applications of Newton-Type Iterations* (Springer, New York, 2008)
10. I.K. Argyros, On a class of Newton-like methods for solving nonlinear equations. J. Comput. Appl. Math. **228**, 115–122 (2009)
11. I.K. Argyros, A semilocal convergence analysis for directional Newton methods. Math. Comput. AMS **80**, 327–343 (2011)
12. I.K. Argyros, Y.J. Cho, S. Hilout, *Numerical Methods for Equations and Its Applications* (CRC Press/Taylor and Fracncis Group, New York, 2012)
13. I.K. Argyros, S. Hilout, Weaker conditions for the convergence of Newton's method. J. Complex. **28**, 364–387 (2012)
14. J.A. Ezquérro, J.M. Gutiérrez, M.A. Hernández, N. Romero, M.J. Rubio, The Newton method: from Newton to Kantorovich (Spanish). Gac. R. Soc. Mat. Esp. **13**, 53–76 (2010)
15. J.A. Ezquérro, M.A. Hernández, Newton-type methods of high order and domains of semilocal and global convergence. Appl. Math. Comput. **214**(1), 142–154 (2009)

16. L.V. Kantorovich, G.P. Akilov, *Functional Analysis in Normed Spaces* (Pergamon Press, New York, 1964)
17. A.A. Magreñán, Different anomalies in a Jarratt family of iterative root finding methods. Appl. Math. Comput. **233**, 29–38 (2014)
18. A.A. Magreñán, A new tool to study real dynamics: the convergence plane. Appl. Math. Comput. **248**, 215–224 (2014)
19. F.A. Potra, V. Ptak, *Nondiscrete Induction and Iterative Processes* (Pitman Publisher, London, 1984)
20. P.D. Proinov, New general convergence theory for iterative processes and its applications to Newton-Kantorovich type theorems. J. Complex. **26**, 3–42 (2010)

Chapter 6
Monotone Convergence of Iterative Schemes and Their Applications in Fractional Calculus

We present monotone convergence results for general iterative methods in order to approximate a solution of a nonlinear equation defined on a partially ordered linear topological space. Some applications are also provided from fractional calculus using Caputo and Canavati type fractional derivatives. It follows [5].

6.1 Introduction

The first convergence results for Newton's method we given by J.B. Fourier in 1818 and A. Cauchy in 1829 [10]. Cauchy's result is semi-local and makes stronger assumptions on the starting point while Fourier's result makes stronger assumptions on the function and it is a global result. Fourier's result was generalized by A.N. Baluve in 1952 [8] for order-convex operators defined in partially ordered topological linear spaces. Numerous other authors have contributed in this area using iterative methods involving linear, Newton-like and secant-like operators. A survey of such results can be found in [6, 7] and the references therein.

In the present chapter we show convergence for the iterative method defined for each $n = 0, 1, 2, \ldots$ implicitly by

$$F(x_n) + A_n(x_{n+1} - x_n) = 0, \tag{6.1.1}$$

where $F : D \subset X \to Y$, X, Y are POTL (partially ordered topological spaces to be precised in Sect. 6.2), D a subset of X and $A_n : D \to Y$ is a sequence of operators not necessarily in $L(X, Y)$ (the space of bounded linear operators from X into Y). It is worth noticing that as far as we know all previous works in this area are such that $A_n \in L(X, Y)$.

We provide sufficient semi-local convergence criteria such that the sequence $\{x_n\}$ generated by method (6.1.1) converges to a solution x^* of the equation

G.A. Anastassiou and I.K. Argyros, *Intelligent Numerical Methods II: Applications to Multivariate Fractional Calculus*, Studies in Computational Intelligence 649, DOI 10.1007/978-3-319-33606-0_6

$$F(x) = 0. \tag{6.1.2}$$

The convergence results can apply to many areas [1–12]. In particular, the results apply to fractional calculus in connection to Caputo and Canavati type fractional order derivatives, where the operator involved (i.e. A_n) does not have to be necessarily a linear map.

The rest of the chapter is organized as follows: Sect. 6.2 contains the semi-local convergence analysis of method (6.1.1). The four applications from fractional calculus can be found in the concluding Sect. 6.3.

6.2 Semi-local Convergence

We present the semi-local convergence analysis of method (6.1.1) in this section. But first in order to make the paper as self contained as possible we state some concepts involving the theory of partially ordered linear spaces. More details can be found in [6, 7, 11] and the references there in.

Let X be a linear space. A subset C of X is called a cone if $C + C \subseteq C$ and $\alpha C \subseteq C$ for $\alpha > 0$. The cone C is proper if $C \cap (-C) = \{0\}$. The relation "$\leq$" defined by

$$x \leq y \quad \text{if and only if} \quad y - x \in C \tag{6.2.1}$$

is a partial ordering on C which is compatible with the linear structure of this space. Two elements x and y of X are called comparable if either $x \leq y$ or $y \leq x$ holds. The space X endowed with the above relation is called a partially ordered linear space (POL-space). If X has a topology compatible with its linear structure and if the cone C is closed in that topology when X is called a partially ordered topological space (POTL-space).

We remark that in a POTL-space the intervals $[a, b] = \{x : a \leq x \leq b\}$ are closed sets. A stronger connection is considered in the following definitions:

Definition 6.1 A POTL-space is called normal if, given a local base V for the topology, there exists a positive number η so that if $0 \leq z \in U \in V$ then $[0, z] \subset \eta U$.

Definition 6.2 A POTL-space is called regular if every order bounded increasing sequence has a limit.

If the topology of a POTL-space is given by a norm then this space is called a partially ordered normed space (PON-space). If a PON-space is complete with respect to its topology then it is called a partially ordered Banach space (POB-space). According to Definition 6.1 a PON-space is normal if and only if there exists a positive number α such that

$$\|x\| \leq \alpha \|y\| \quad \text{for all } x, y \in X \quad \text{with } 0 \leq x \leq y. \tag{6.2.2}$$

Let us note that any regular POB-space is normal. The reverse is not true. For example, the space $C[0, 1]$ of all continuous real functions defined on $[0, 1]$, ordered by the cone of nonnegative functions, is normal but is not regular. All finite dimensional POTL-spaces are both normal and regular.

We state a well known result of Kantorovich which will be used in the proof of our main theorem that follows [9].

Proposition 6.3 *Let X be a regular POTL-space and let x, y be two points of X such that $x \leq y$. If $P : [x, y] \to X$ is a continuous isotone operator having the property that $x \leq Px$ and $y \geq Py$, then there exists a point $z \in [x, y]$ such that $z = Pz$.*

Next, we present the main monotone convergence result for method (6.1.1).

Theorem 6.4 *Let $F : D \subseteq X \to Y$, where X is a regular POTL-space and Y is a POTL-space. Let x_0, y_0, y_{-1} be three points of D such that*

$$x_0 \leq y_0 \leq y_{-1}, \quad \left[x_0, y_{-1}\right] \subset D, \quad Fx_0 \leq 0 \leq Fy_0. \tag{6.2.3}$$

Set:

$$\begin{aligned} &D_1 = \left\{(x, y) \in X^2 : x_0 \leq x \leq y \leq y_0\right\}, \\ &D_2 = \left\{(u, y_{-1}) \in X^2 : x_0 \leq u \leq y_0\right\}, \\ &\text{and} \\ &D_3 = D_1 \cup D_2. \end{aligned} \tag{6.2.4}$$

Suppose that there exist a mapping $A : D_3 \to Y$ such that

$$Fy - Fx \leq A(w, z)(y - x) \tag{6.2.5}$$

and

$$0 \leq A(w, z)\, y - A(w, z)\, x \leq y - x, \tag{6.2.6}$$

for all $(x, y), (y, w) \in D_1$, $(w, z) \in D_3$.

Then, there exist two sequences $(x_n)_{n\geq 1}$, $(y_n)_{n\geq 1}$ and two points x^*, y^* of X satisfying for each $n = 0, 1, 2, \ldots$:

$$Fy_n + A(y_n, y_{n-1})(y_{n+1} - y_n) = 0 \tag{6.2.7}$$

$$Fx_n + A(y_n, y_{n-1})(x_{n+1} - x_n) = 0 \tag{6.2.8}$$

$$Fx_n \leq 0 \leq Fy_n \tag{6.2.9}$$

$$x_0 \leq x_1 \leq \cdots \leq x_n \leq x_{n+1} \leq y_{n+1} \leq y_n \leq \cdots \leq y_1 \leq y_0 \tag{6.2.10}$$

$$\lim_{n\to\infty} x_n = x^*, \quad \lim_{n\to\infty} y_n = y^*. \tag{6.2.11}$$

Moreover, if operator F is continuous and $x^* = y^*$. Then, we have that $F(x^*) = 0$.

Proof Let operator $P : [0, y_0 - x_0] \to X$ be defined by

$$Px = x - (Fx_0 + A_0 x). \tag{6.2.12}$$

It is easy to see by (6.2.6) and (6.2.12) that P is isotone and continuous. We also have

$$\begin{aligned} P(0) &= -Fx_0 \geq 0 \\ P(y_0 - x_0) &= y_0 - x_0 - Fy_0 + (Fy_0 - Fx_0 - A_0(y_0 - x_0)) \\ &\leq y_0 - x_0 - Fy_0 \leq y_0 - x_0. \end{aligned}$$

In view of Proposition 6.3, the operator P has a fixed point $w \in [0, y_0 - x_0]$. Let $x_1 = x_0 + w$. Then, we have that

$$Fx_0 + A_0(x_1 - x_0) = 0, \quad x_0 \leq x_1 \leq y_0.$$

Using (6.2.5) from the preceding relations, we have that

$$Fx_1 = Fx_1 - Fx_0 + A_0(x_0 - x_1) \leq 0.$$

Define $G : [0, y_0 - x_1] \to P$ by

$$Gx = x + (Fy_0 - A_0 x).$$

It follows that G is clearly continuous and isotone. We get that

$$\begin{aligned} G(0) &= Fy_0 \geq 0 \\ G(y_0 - x_1) &= y_0 - x_1 + Fx_1 + (Fy_0 - Fx_1 - A_0(y_0 - x_1)) \\ &\leq y_0 - x_1 + Fx_1 \leq y_0 - x_1. \end{aligned}$$

Then, again by Proposition 6.3, there exists $z \in [0, y_0 - x_1]$ such that $z = Gz$. Let $y_1 = y_0 - z$. It follows that

$$Fy_0 + A(y_1 - y_0) = 0, \quad x_1 \leq y_1 \leq y_0.$$

Using the above relations and condition (6.2.6), we get that

$$Fy_1 = Fy_1 - Fy_0 - A_0(y_1 - y_0) \geq 0.$$

Proceeding by induction we can show that there exist two sequences $(x_n)_{n\geq 1}$ and $(y_n)_{n\geq 1}$ satisfying (6.2.7)–(6.2.10) in a regular space X and as such it follows from (6.2.10) that there exist $x^*, y^* \in X$ such that $x^* = \lim_{n\to\infty} x_n$, $y^* = \lim_{n\to\infty} y_n$.

We obviously have $x^* \leq y^*$. Finally, using the continuity of F and (6.2.9), we have that $F(x^*) \leq 0 \leq F(y^*)$. Hence, we conclude that $F(x^*) = 0$.

Remark 6.5 If $A : D_3 \to L(X, Y)$, then Theorem 6.4 reduces to earlier results [6–11].

6.3 Applications to Fractional Calculus

I. We need:

Definition 6.6 Let $0 < \alpha < 1$, $f : [a, b] \to \mathbb{R}$, $f' \in L_\infty([a, b])$, and $y \geq x$; $x, y \in [a, b]$. The left Caputo fractional derivative of f is defined by

$$\left(D_{x+}^{\alpha} f\right)(y) = \frac{1}{\Gamma(1-\alpha)} \int_x^y (y-t)^{-\alpha} f'(t)\, dt, \tag{6.3.1}$$

where Γ is the gamma function.

By convention $D_{x+}^{1} f = f'$ and $\left(D_{x+}^{\alpha} f\right)(y) = 0$, for $y < x$.

Clearly $D_{x+}^{\alpha} f$ is a continuous function over $[a, b]$, see [1], p. 388.

Notice that if f is decreasing, then f is differentiable a.e. and $f' \leq 0$ [12], and $\left(D_{x+}^{\alpha} f\right)(y) \leq 0$, $\forall\, y \in [a, b]$.

From [4], we mention:

Theorem 6.7 (left fractional mean value theorem) *Let $0 < \alpha < 1$ and $f \in AC([a, b])$ (absolutely continuous function), $f' \in L_\infty([a, b])$. Then*

$$f(x) - f(a) = \frac{\left(D_{a+}^{\alpha} f\right)(\xi_x)}{\Gamma(\alpha+1)} (x-a)^{\alpha}, \tag{6.3.2}$$

where $\xi_x \in [a, x]$, any $x \in [a, b]$.

Assumption 6.8 *In this application we consider $0 < \alpha < 1$, $f \in AC([a, b])$, $f' \in L_\infty([a, b])$, f is decreasing; $f(b) < 0 < f(a)$. We consider any $x, y \in [a, b]$:*

$$0 \leq y - x \leq 1. \tag{6.3.3}$$

By Theorem 6.7 we get that

$$f(y) - f(x) = \frac{\left(D_{x+}^{\alpha} f\right)(\xi_{xy})}{\Gamma(\alpha+1)} (y-x)^{\alpha}, \tag{6.3.4}$$

where $\xi_{xy} \in [x, y]$.

Notice that

$$(y-x)^{\alpha} \geq (y-x). \tag{6.3.5}$$

Clearly, from the above (6.3.4), we can have that

$$f(y) - f(x) \leq \frac{\left(D_{x+}^{\alpha} f\right)\left(\xi_{xy}\right)}{\Gamma(\alpha+1)} (y-x). \tag{6.3.6}$$

Now, we can apply our numerical method (6.1.1) to solve

$$f(x) = 0, \tag{6.3.7}$$

with A defined by

$$A(x, y) = \frac{\left(D_{x+}^{\alpha} f\right)\left(\xi_{xy}\right)}{\Gamma(\alpha+1)}.$$

II. We need:

Definition 6.9 Let $0 < \alpha < 1$, $f : [a, b] \to \mathbb{R}$, $f' \in L_{\infty}([a, b])$, and $y \geq x$; $x, y \in [a, b]$. The right Caputo fractional derivative of f is defined by

$$\left(D_{y-}^{\alpha} f\right)(x) = \frac{-1}{\Gamma(1-\alpha)} \int_{x}^{y} (t-x)^{-\alpha} f'(t)\, dt. \tag{6.3.8}$$

Notice that $D_{y-}^{1} f = -f'$, by convention. Clearly here $D_{y-}^{\alpha} f \in C([a, b])$, see [2]. We make the convention that $\left(D_{y-}^{\alpha} f\right)(x) = 0$, for $x > y$.

Notice that if f is decreasing, then f is differentiable a.e. and $f' \leq 0$, and $\left(D_{y-}^{\alpha} f\right)(x) \geq 0$, $\forall\, x \in [a, b]$.

From [4], we mention:

Theorem 6.10 (right fractional mean value theorem) *Let* $0 < \alpha < 1$, $f \in AC([a, b])$, $f' \in L_{\infty}([a, b])$. *Then*

$$f(x) - f(b) = \frac{\left(D_{b-}^{\alpha} f\right)(\xi_x)}{\Gamma(\alpha+1)} (b-x)^{\alpha}, \tag{6.3.9}$$

with $x \leq \xi_x \leq b$, *where* $x \in [a, b]$.

Still, here we suppose Assumption 6.8. By Theorem 6.10, we get that

$$f(x) - f(y) = \frac{\left(D_{y-}^{\alpha} f\right)\left(\xi_{xy}\right)}{\Gamma(\alpha+1)} (y-x)^{\alpha}, \tag{6.3.10}$$

where $\xi_{xy} \in [x, y]$.

Hence, it holds (by (6.3.10))

$$f(y) - f(x) = \frac{-\left(D_{y-}^{\alpha} f\right)(\xi_{xy})}{\Gamma(\alpha+1)} (y-x)^{\alpha} \overset{\text{(by (6.3.5))}}{\leq} \frac{-\left(D_{y-}^{\alpha} f\right)(\xi_{xy})}{\Gamma(\alpha+1)} (y-x). \tag{6.3.11}$$

Now, we can solve numerically

$$f(x) = 0, \tag{6.3.12}$$

with A defined by

$$A(x, y) = -\frac{\left(D_{y-}^{\alpha} f\right)(\xi_{xy})}{\Gamma(\alpha+1)}.$$

III. See [1], pp. 7–8.

Let $\nu > 0$, $f \in C([a, b])$, we define the left Riemann–Liouville fractional integral as

$$\left(J_{\nu}^{z_0} f\right)(z) = \frac{1}{\Gamma(\nu)} \int_{z_0}^{z} (z-t)^{\nu-1} f(t)\, dt, \tag{6.3.13}$$

for $a \leq z_0 \leq z \leq b$.

We set $J_0^{z_0} f = f$. Let here $1 \leq \nu < 2$. Let $[\nu]$ (the integral part of ν) and $\alpha = \nu - [\nu]$ $(0 < \alpha < 1)$, here it is $\alpha = \nu - 1$. We define the subspace $C_{x_0}^{\nu}([a, b])$ of $C^1([a, b])$, where $x_0 \in [a, b]$:

$$C_{x_0}^{\nu}([a, b]) := \left\{ f \in C^1([a, b]) : J_{1-\alpha}^{x_0} f' \in C^1([x_0, b]) \right\}. \tag{6.3.14}$$

So let $f \in C_{x_0}^{\nu}([a, b])$; we define the left Canavati type fractional derivative of f of order ν, over $[x_0, b]$ as

$$D_{x_0}^{\nu} f = \left(J_{1-\alpha}^{x_0} f'\right)'. \tag{6.3.15}$$

Clearly, for $f \in C_{x_0}^{\nu}([a, b])$, there exists and it is continuous,

$$\left(D_{x_0}^{\nu} f\right)(z) = \frac{1}{\Gamma(1-\alpha)} \frac{d}{dz} \int_{x_0}^{z} (z-t)^{-\alpha} f'(t)\, dt, \tag{6.3.16}$$

for all $x_0 \leq z \leq b$.

In [1], p. 8 we have proved the fundamental theorem of fractional calculus, that follows:

Theorem 6.11 *Let $f \in C_{x_0}^{\nu}([a, b])$, $x_0 \in [a, b]$ is fixed, $1 \leq \nu < 2$. Then*

$$f(x) - f(x_0) = \frac{1}{\Gamma(\nu)} \int_{x_0}^{x} (x-t)^{\nu-1} \left(D_{x_0}^{\nu} f\right)(t)\, dt, \tag{6.3.17}$$

all $x \in [a, b] : x \geq x_0$.

By the first integral mean value theorem, and (6.3.17), we get that

$$f(x) - f(x_0) = \frac{\left(D_{x_0}^{\nu} f\right)\left(\xi_{x_0 x}\right)}{\Gamma(\nu)} \int_{x_0}^{x} (x-t)^{\nu-1} dt$$
$$= \frac{\left(D_{x_0}^{\nu} f\right)\left(\xi_{x_0 x}\right)}{\Gamma(\nu)} \frac{(x-x_0)^{\nu}}{\nu} = \frac{\left(D_{x_0}^{\nu} f\right)\left(\xi_{x_0 x}\right)}{\Gamma(\nu+1)} (x-x_0)^{\nu}, \quad (6.3.18)$$

where $\xi_{x_0 x} \in [x_0, x]$.

We have proved:

Theorem 6.12 (Mean value theorem of left fractional calculus) *Let $f \in C_{x_0}^{\nu}([a,b])$, $x_0 \in [a,b]$ is fixed, $1 \leq \nu < 2$. Then*

$$f(x) - f(x_0) = \frac{\left(D_{x_0}^{\nu} f\right)\left(\xi_{x_0 x}\right)}{\Gamma(\nu+1)} (x-x_0)^{\nu}, \quad (6.3.19)$$

where $\xi_{x_0 x} \in [x_0, x]$, for any $x \in [a,b] : x \geq x_0$.

Our mathematics for our numerical model follow:

Here, we consider $f \in C^1([a,b])$, such that $f \in C_x^{\nu}([a,b])$, $1 \leq \nu < 2$, for every $x \in [a,b]$, we write that as $f \in C_+^{\nu}([a,b])$. So we only consider next $f \in C_+^{\nu}([a,b])$. Thus, (6.3.19) reads now as follows:

$$f(y) - f(x) = \frac{\left(D_x^{\nu} f\right)\left(\xi_{xy}\right)}{\Gamma(\nu+1)} (y-x)^{\nu}, \quad (6.3.20)$$

$\forall\, x, y \in [a,b] : y \geq x$.

We further consider $x, y \in [a,b]$ such that $|y-x| \leq 1$, and that f is increasing, and $f(a) < 0 < f(b)$. See that

$$(y-x)^{\nu} \leq (y-x), \quad 1 \leq \nu < 2, \; y \geq x.$$

Hence by (6.3.20) we obtain

$$f(y) - f(x) \leq \frac{\left|\left(D_x^{\nu} f\right)\left(\xi_{xy}\right)\right|}{\Gamma(\nu+1)} (y-x), \quad (6.3.21)$$

$\forall\, x, y \in [a,b] : y \geq x$.

One may consider $f : D_x^{\nu} f \geq 0$, for any $x \in [a,b]$, then we can have

$$f(y) - f(x) \leq \frac{\left(D_x^{\nu} f\right)\left(\xi_{xy}\right)}{\Gamma(\nu+1)} (y-x), \quad (6.3.22)$$

$\forall\, x, y \in [a,b] : y \geq x$.

Here we take as

$$A(x, y) = \frac{\left|\left(D_x^{\nu} f\right)(\xi_{xy})\right|}{\Gamma(\nu + 1)}.$$

IV. From [2], pp. 345–348, we mention the following background:

Let $\nu > 0$, $n := [\nu]$, $\alpha = \nu - n$, $0 < \alpha < 1$, $f \in C([a, b])$, call the right Riemann–Liouville fractional integral operator by

$$\left(J_{b-}^{\nu} f\right)(x) := \frac{1}{\Gamma(\nu)} \int_x^b (\zeta - x)^{\nu - 1} f(\zeta)\, d\zeta, \tag{6.3.23}$$

$x \in [a, b]$. Define the subspace of functions

$$C_{b-}^{\nu}([a, b]) := \left\{ f \in C^n([a, b]) : J_{b-}^{1-\alpha} f^{(n)} \in C^1([a, b]) \right\}. \tag{6.3.24}$$

Define the right generalized ν-fractional derivative of f over $[a, b]$ as

$$D_{b-}^{\nu} f := (-1)^{n-1} D J_{b-}^{1-\alpha} f^{(n)}, \quad D = \frac{d}{dx}. \tag{6.3.25}$$

Notice that

$$J_{b-}^{1-\alpha} f^{(n)}(x) = \frac{1}{\Gamma(1-\alpha)} \int_x^b (\zeta - x)^{-\alpha} f^{(n)}(\zeta)\, d\zeta \tag{6.3.26}$$

exists for $f \in C_{b-}^{\nu}([a, b])$, and

$$D_{b-}^{\nu} f(x) = \frac{(-1)^{n-1}}{\Gamma(1-\alpha)} \frac{d}{dx} \int_x^b (\zeta - x)^{-\alpha} f^{(n)}(\zeta)\, d\zeta. \tag{6.3.27}$$

I.e.

$$\left(D_{b-}^{\nu} f\right)(x) = \frac{(-1)^{n-1}}{\Gamma(n - \nu + 1)} \frac{d}{dx} \int_x^b (\zeta - x)^{n-\nu} f^{(n)}(\zeta)\, d\zeta. \tag{6.3.28}$$

If $\nu \in \mathbb{N}$, then $\alpha = 0$, $n = \nu$, and

$$D_{b-}^{\nu} f(x) = (-1)^n f^{(n)}(x). \tag{6.3.29}$$

We mention the following Taylor fractional formulae ([2], p. 348)

Theorem 6.13 *Let $f \in C_{b-}^{\nu}([a,b])$, $\nu > 0$, $n := [\nu]$. Then*

1. *If $\nu \geq 1$, we get*

$$f(x) = \sum_{k=0}^{n-1} \frac{f^{(k)}(b_-)}{k!}(x-b)^k + \left(J_{b-}^{\nu} D_{b-}^{\nu} f\right)(x), \quad \forall x \in [a,b]. \tag{6.3.30}$$

2. *If $0 < \nu < 1$, we get*

$$f(x) = J_{b-}^{\nu} D_{b-}^{\nu} f(x), \quad \forall x \in [a,b]. \tag{6.3.31}$$

From now on we consider $1 \leq \nu < 2$. Hence (6.3.30) implies

$$f(x) - f(b) = \left(J_{b-}^{\nu} D_{b-}^{\nu} f\right)(x) =$$

$$\frac{1}{\Gamma(\nu)} \int_x^b (z-x)^{\nu-1} \left(D_{b-}^{\nu} f\right)(z)\, dz = \frac{\left(D_{b-}^{\nu} f\right)(\xi_{xb})}{\Gamma(\nu+1)}(b-x)^{\nu}, \quad \forall x \in [a,b], \tag{6.3.32}$$

with $\xi_{xb} \in [x,b]$.

We have proved:

Theorem 6.14 (Mean value theorem of right fractional calculus) *Let $f \in C_{b-}^{\nu}([a,b])$, $1 \leq \nu < 2$. Then*

$$f(x) - f(b) = \frac{\left(D_{b-}^{\nu} f\right)(\xi_{xb})}{\Gamma(\nu+1)}(b-x)^{\nu}, \tag{6.3.33}$$

$\forall\, x \in [a,b]$, *where $\xi_{xb} \in [a,b]$.*

Our mathematics for our numerical model follow:

Here, we consider $f \in C^1([a,b])$, such that $f \in C_{y-}^{\nu}([a,b])$, $1 \leq \nu < 2$, for every $y \in [a,b]$, we write that as $f \in C_{-}^{\nu}([a,b])$. So we only consider next $f \in C_{-}^{\nu}([a,b])$, with $1 \leq \nu < 2$. For convenience we state that

$$C_{y-}^{\nu}([a,b]) = \left\{f \in C^1([a,b]) : J_{y-}^{1-\alpha} f' \in C^1([a,y])\right\}, \quad y \in [a,b], \tag{6.3.34}$$

which the same as (6.3.24), when $y = b$, for $1 \leq \nu < 2$.

Thus, (6.3.33) reads now as follows:

$$f(x) - f(y) = \frac{\left(D_{y-}^{\nu} f\right)(\xi_{xy})}{\Gamma(\nu+1)}(y-x)^{\nu}, \tag{6.3.35}$$

$\forall\, x \in [a,y]$, $y \in [a,b]$, where $\xi_{xy} \in [a,y]$.

We further consider $x, y \in [a,b]$ such that $|y-x| \leq 1$, and that f is increasing, and $f(a) < 0 < f(b)$. See that $(y-x)^{\nu} \leq (y-x)$, $1 \leq \nu < 2$, $y \geq x$. Hence by (6.3.35) we obtain

$$|f(x) - f(y)| = \frac{\left|\left(D_{y-}^{\nu} f\right)(\xi_{xy})\right|}{\Gamma(\nu+1)} |y-x|^{\nu} \leq \frac{\left|\left(D_{y-}^{\nu} f\right)(\xi_{xy})\right|}{\Gamma(\nu+1)} (y-x), \quad y \geq x. \tag{6.3.36}$$

That is, we derive:

$$f(y) - f(x) \leq \frac{\left|\left(D_{y-}^{\nu} f\right)(\xi_{xy})\right|}{\Gamma(\nu+1)} (y-x), \tag{6.3.37}$$

$\forall\, x \in [a, y]$, $y \in [a, b]$, with $\xi_{xy} \in [a, b]$, $1 \leq \nu < 2$.

One may consider $f : D_{y-}^{\nu} f \geq 0$, for every $y \in [a, b]$, then we can have

$$f(y) - f(x) \leq \frac{\left(D_{y-}^{\nu} f\right)(\xi_{xy})}{\Gamma(\nu+1)} (y-x), \tag{6.3.38}$$

$\forall\, x \in [a, y]$, $y \in [a, b]$, where $\xi_{xy} \in [a, b]$, $1 \leq \nu < 2$.

Here we take A as follows:

$$A(x, y) = \frac{\left|\left(D_{y-}^{\nu} f\right)(\xi_{xy})\right|}{\Gamma(\nu+1)}.$$

References

1. G. Anastassiou, *Fractional Differentiation Inequalities* (Springer, New York, 2009)
2. G. Anastassiou, *Inteligent Mathematics: Computational Analysis* (Springer, Heidelberg, 2011)
3. G. Anastassiou, Fractional representation formulae and right fractional inequalities. Math. Comput. Model. **54**(10–12), 3098–3115 (2011)
4. G. Anastassiou, Advanced fractional Taylor's formulae. J. Comput. Anal. Appl. **21**(7), 1185–1204 (2016)
5. G. Anastassiou, I. Argyros, *On the Monotone Convergence of General Iterative Methods with Applications in Fractional Calculus* (2015), submitted for publication
6. I.K. Argyros, F. Szidarovszky, in *The Theory and Applications of Iteration Methods*, Systems Engineering Series, ed. by A. Terry Bahill (CRC Press, Boca Raton, 1993)
7. I.K. Argyros, in *Computational Theory of Iterative methods*, ed. by C.K. Chvi, L. Wvytack. Studies in Computational Mathematics, vol. 15 (Elsevier Publ. Co., New York, 2007)
8. A.N. Baluev, On the abstract theory of Chaplygin's method, (Russian). Dokl. Akad. Nauk. SSSR **83**, 781–784 (1952)
9. L.V. Kantorovitch, The method of succesive approximation for functional equations. Acta Math. **71**, 63–97 (1939)
10. J.M. Ortega, W.C. Rheinboldt, *Iterative Solution of Nonlinear Equations in Several Variables* (Academic Press, New York, 1970)
11. J.S. Vandergraft, Newton's method for convex operators in partially ordered spaces. SIAM J. Numer. Anal. **4**, 406–432 (1967)
12. A.R. Schep, *Differentiation of Monotone Functions*,https://people.math.sc.edu/schep/diffmonotone.pdf

Chapter 7
Extending the Convergence Domain of Newton's Method

We present a local as well a semilocal convergence analysis for Newton's method in a Banach space setting. Using the same Lipschitz constants as in earlier studies [2, 4–9, 11, 13–15] we extend the applicability of, Newton's method as follows: Local case: A larger radius is given as well as more precise error estimates on the, distances involved. Semilocal case: the convergence domain is extended; the error estimates are tighter and the information on the location of the solution is at least as precise as before. Numerical examples further justify the theoretical results. It follows [10].

7.1 Introduction

In this chapter we are concerned with the problem of approximating a locally unique solution x^* of equation

$$F(x) = 0, \tag{7.1.1}$$

where F is a Fréchet-differentiable operator defined an open convex subset D of a Banach space X with values in a Banach space Y.

Many problems from Applied Sciences including engineering can be solved by means of finding the solutions of equations in a form like (7.1.1) using Mathematical Modelling [1, 2, 5, 8, 13, 15]. Except in special cases, the solutions of these equations can be found in closed form. This is the main reason why the most commonly used solution methods are usually iterative. The convergence analysis of iterative methods is usually divided into two categories: semilocal and local convergence analysis. The semilocal convergence matter is, based on the information around an initial point, to give criteria ensuring the convergence of iteration procedures. A very important problem in the study of iterative procedures is the convergence domain. In general the convergence domain is small. Therefore, it is important to enlarge the convergence domain without additional hypothesis. Another important problem is to find more precise error estimates on the distances $\|x_{n+1} - x_n\|$, $\|x_n - x^*\|$.

G.A. Anastassiou and I.K. Argyros, *Intelligent Numerical Methods II: Applications to Multivariate Fractional Calculus*, Studies in Computational Intelligence 649, DOI 10.1007/978-3-319-33606-0_7

Newton's method defined for each $n = 0, 1, 2, \ldots$ by

$$x_{n+1} = x_n - F'(x_n)^{-1}F(x_n), \tag{7.1.2}$$

where x_0 is an initial point,is undoubtedly the most popular method for generating a sequence $\{x_n\}$ approximating x^*. There is a plethora on local as well as semilocal convergence results for Newton's method [1–9, 11–13, 15]. The conditions (C) for the semilocal convergence are:

(C_1) $F : D \subset X \to Y$ is Fréchet differentiable and there exist $x_0 \in D$, $\eta \geq 0$ such that $F'(x_0)^{-1} \in L(Y, X)$ and

$$\|F'(x_0)^{-1}F(x_0)\| \leq \eta$$

(C_2) There exists a parameter $L > 0$ such that for each $x, y \in D$

$$\|F'(x_0)^{-1}(F'(x) - F'(y))\| \leq L\|x - y\|.$$

In view of (C_2):

(C_3) There exists $L_0 > 0$ such that

$$\|F'(x_0)^{-1}(F'(x) - F'(x_0))\| \leq L_0\|x - x_0\|.$$

Clearly, we have that

$$L_0 \leq L \tag{7.1.3}$$

and $\frac{L}{L_0}$ can be arbitrarily large [4]. It is worth noticing that (7.1.3) is not an additional to (7.1.2) hypothesis, since in practice the computation of constant L involves the computation of constant L_0 as a special case.

Let $U(z, \varrho)$, $\bar{U}(z, \varrho)$ stand, respectively for the open and closed ball in X with center $z \in X$ and radius $\varrho > 0$.

The sufficient convergence criteria for Newton's method using the conditions (C), constants L, L_0 and η given in affine invariant form are:

- Kantorovich [13]

$$h_K = 2L\eta \leq 1. \tag{7.1.4}$$

- Argyros [4]

$$h_1 = (L_0 + L)\eta \leq 1. \tag{7.1.5}$$

- Argyros [5]

$$h_2 = \frac{1}{4}\left(L + 4L_0 + \sqrt{L^2 + 8L_0L}\right)\eta \leq 1 \tag{7.1.6}$$

- Argyros [7, 8]

$$h_3 = \frac{1}{4}\left(4L_0 + \sqrt{L_0 L + 8L_0^2} + \sqrt{L_0 L}\right)\eta \leq 1 \tag{7.1.7}$$

If $L_0 = L$, then (7.1.5)–(7.1.7) coincide with (7.1.4). If $L_0 < L$, then

$$h_K \leq 1 \Rightarrow h_1 \leq 1 \Rightarrow h_2 \leq 1 \Rightarrow h_3 \leq 1,$$

but not vice versa. We also have that

$$\begin{aligned} &\frac{h_1}{h_K} \to \frac{1}{2}, \quad \frac{h_2}{h_K} \to \frac{1}{4}, \quad \frac{h_2}{h_1} \to \frac{1}{2} \\ &\frac{h_3}{h_K} \to 0, \quad \frac{h_3}{h_1} \to 0, \quad \frac{h_3}{h_2} \to 0 \\ &\text{as } \frac{L_0}{L} \to 0. \end{aligned} \tag{7.1.8}$$

Conditions (7.1.8) show by how many times (at most) the better condition improves the less better condition. Therefore the condition to improve is (7.1.7). This is done as follows: Replace condition (C_2) by

$(C_2)'$ There exists $L_1 > 0$ such that

$$\|F'(x_0)^{-1}(F'(x) - F'(y))\| \leq L_1 \|x - y\|.$$

for each $x, y \in U\left(x_0, \frac{1}{L_0}\right) \cap D$.

Denote the conditions (C_1), (C_3) and $(C_2)'$ by $(C)'$. Clearly, we have that

$$L_1 \leq L \tag{7.1.9}$$

holds in general.

Then, by simply replacing (C_2) by $(C_2)'$ and using the conditions $(C)'$ instead of using the conditions (C) we can replace the L by L_1 in all convergence criteria (7.1.5)–(7.1.7). In particular we have that

$$h_4 = \frac{1}{4}\left(4L_0 + \sqrt{L_0 L_1 + 8L_0^2} + \sqrt{L_0 L_1}\right)\eta \leq 1 \tag{7.1.10}$$

In view of (7.1.7), (7.1.9) and (7.1.10) we get that

$$h_3 \leq 1 \Rightarrow h_4 \leq 1. \tag{7.1.11}$$

but not necessarily vice versa unless if $L_1 = L$.

Notice that there is an even weaker condition that (7.1.7) in [9]. This condition uses L_0, L and η and some other constants that can be smaller than L_0, L.

Similarly the conditions (H) for the local convergence are:

(H_1) $F : D \subset X \to Y$ is Fréchet differentiable and there exist $x^* \in D$, such that $F(x^*)^{-1} \in L(Y, X)$

(H_2) There exists a parameter $K > 0$ such that for each $x, y \in D$

$$\|F'^*)^{-1}(F'(x) - F'(y))\| \leq K\|x - y\|.$$

(H_3) There exists a parameter $K_0 > 0$ such that for each $x, y \in D$

$$\|F'^*)^{-1}(F'(x) - F'^*))\| \leq K_0\|x - x_0\|.$$

We have that

$$K_0 \leq K \tag{7.1.12}$$

and $\frac{K}{K_0}$ can be arbitrarily large [4]. The radius of convergence is:

- Rheinboldt [14], Traub [15]

$$R_0 = \frac{2}{3K} \tag{7.1.13}$$

- Argyros [4]

$$R_1 = \frac{2}{2K_0 + K}. \tag{7.1.14}$$

In view of (7.1.12)–(7.1.14) for $K_0 < K$

$$R_0 < R_1 \tag{7.1.15}$$

and

$$\frac{R_1}{R_0} \to \frac{1}{3} \quad \text{as} \quad \frac{K_0}{K} \to 0. \tag{7.1.16}$$

Estimate (7.1.16) shows that R_1 is at most three times larger than R_0. Therefore the radius to improve is (7.1.14). Replace (H_2) by

$(H_2)'$ There exists $K_1 > 0$ such that for each $x, y \in U\left(x^*, \frac{1}{K_0}\right) \cap D$

$$\|F'^*)^{-1}\left(F'(x) - F'(y)\right)\| \leq K_1\|x - y\|.$$

Denote by $(H)'$ conditions (H_1), (H_3) and $(H_2)'$. Then, we obtain the radius of convergence

$$R_2 = \frac{2}{2K_0 + K_1}. \tag{7.1.17}$$

In view of $(H_2)'$ and (H_2) we have

$$K_1 \leq K \tag{7.1.18}$$

and consequently for $K_1 < K$

$$R_1 < R_2. \tag{7.1.19}$$

The chapter is organized as follows. In Sect. 7.2 we present the semilocal as well as the local convergence analysis of Newton's method (7.1.2). The numerical examples are presented in the concluding Sect. 7.3.

7.2 Convergence Analysis

We present first the semilocal convergence analysis of Newton's method under the $(C)'$ conditions and secondly the local convergence under the $(H)'$ conditions.

It is convenient for the semilocal convergence analysis that follows to introduce the scalar majorizing sequence $\{s_n\}$ defined by

$$\begin{aligned} s_0 = 0, \quad s_1 = \eta, \quad s_2 = s_1 + \frac{L_0(s_1 - s_0)^2}{2(1 - L_0 s_1)} \\ s_{n+2} = s_{n+1} + \frac{L_1(s_{n+1} - s_m)^2}{2(1 - L_0 s_{n+1})}, \quad n = 1, 2, 3, \ldots \end{aligned} \tag{7.2.1}$$

Let $s^* = \lim_{n \to \infty} s_n$.

Then, we state the semilocal convergence results for Newton's method.

Theorem 7.1 *Suppose that the $(C)'$ conditions are satisfied and $\bar{U}(x_0, s^*) \subseteq D$. Then,*

(a) the sequence $\{s_n\}$ generated by (7.2.1) is nondecreasing, bounded above by s^{**} given by

$$s^{**} = \eta + \frac{L_0 \eta^2}{2(1 - \delta)(1 - L_0 \eta)},$$

where

$$\delta = \frac{2L_1}{L_1 + \sqrt{L_1^2 + 8L_0 L_1}}$$

and converges to its unique upper bound s^* which satisfies

$$\eta \leq s^* \leq s^{**}.$$

(b) The sequence $\{x_n\}$ generated by Newton's method (7.1.2) is well defined, remains in $\bar{U}(x_0, s^*)$ and converges to a solution $x^* \in \bar{U}(x_0, s^*)$ of equation $F(x) = 0$.

Moreover, the following estimates hold

$$\|x_{n+1} - x_n\| \leq s_{n+1} - s_n,$$

and

$$\|x_n - x^*\| \leq s_n - s^*.$$

Furthermore, for $\bar{s}^* \geq s^*$ *such that*

$$L_0(\bar{s}^* + s^*) < 2$$

the limit point x^* *is the only solution of equation* $F(x) = 0$ *in* $\bar{U}(x_0, \bar{s}^*)$.

Proof Simply follow the proof of Lemma 2.1 for part (a) and the proof of Theorem 3.2 for part (b) and replace L, $\{t_n\}$ by L_1 and $\{s_n\}$ in [7]. Notice also that the iterates $\{x_n\} \in \bar{U}(x_0, \frac{1}{L_0}) \cap D$. ■

Remark 7.2 (a) The majorizing sequence $\{t_n\}$, t^*, t^{**} given in [7] under conditions (C) and (7.1.7) are defined by

$$\begin{aligned} t_0 = 0, \quad t_1 = \eta, \quad t_2 = t_1 + \frac{L(t_1 - t_0)^2}{2(1 - L_0 t_1)} \\ t_{n+2} = t_{n+1} + \frac{L(t_{n+1} - t_n)^2}{2(1 - L_0 t_{n+1})}, \quad n = 1, 2, \ldots \\ t^* = \lim_{n\to\infty} t_n \leq t^{**} = \eta + \frac{L_0 \eta^2}{2(1 - \delta_0)(1 - L_0 \eta)}, \end{aligned} \tag{7.2.2}$$

where

$$\delta_0 = \frac{2L}{2 + \sqrt{L^2 + 8L_0 L}}.$$

Using a simple inductive argument, (7.2.1) and (7.2.2) we get for $L_1 < L$ and $n = 2, 3, \ldots$ that

$$s_n < t_n, \tag{7.2.3}$$

$$x_{n+1} - s_n < t_{n+1} - t_n \tag{7.2.4}$$

and

$$s^* \leq t^* \tag{7.2.5}$$

Estimates (7.2.3)–(7.2.5) to show the new error bounds are more precise than the old ones and the information on the location of the solution x^* is at least as precise as already claimed in the abstract of this study.

(b) Condition $\bar{U}(x_0, s^*) \subseteq D$ can be replaced by $U(x_0, \frac{1}{L_0})$. In this case condition $(C_2)'$ holds for all $x, y \in U(x_0, \frac{1}{L_0})$.

Next, we present the local convergence analysis of Newton's method (7.1.2) under the $(H)'$ conditions.

Theorem 7.3 *Suppose that the conditions $(H)'$ hold and $\bar{U}(x^*, R_2) \subseteq D$. Then, sequence $\{x_n\}$ generated by Newton's method (7.1.2) for $x_0 \in U(x^*, R_2) \setminus \{x^*\}$ is well defined, remains in $U(x^*, R)$ and converges to x^*. Moreover, the following estimates hold*

$$\|x_{n+1} - x_n\| \leq \frac{\bar{K}_1 \|x_n - x^*\|^2}{2(1 - K_0\|x_n - x^*\|)}, \tag{7.2.6}$$

where

$$\bar{K}_1 = \begin{cases} K_0, \ n = 0 \\ K_1, \ n > 0. \end{cases}$$

Furthermore, for $T \in [R_2, \frac{2}{K_0})$, the limit point x^* is the only solution of equation $F(x) = 0$ in $\bar{U}(x^*, T)$.

Proof Simply repeat the proof of Theorem 3.5 in [7] with K_1, $(H)'$, replacing K, (H), respectively. ■

Remark 7.4 The error bounds given in [4, 7] are given by

$$\|x_{n+1} - x^*\| \leq \frac{\bar{K} \|x_n - x^*\|^2}{2(1 - L_0\|x_n - x^*\|)}, \tag{7.2.7}$$

where

$$\bar{K} = \begin{cases} K_0, \ n = 0 \\ K, \ \ n > 0. \end{cases}$$

In view of (7.2.6) and (7.2.7) we see that if $K_1 < K$, then the new error bounds given by (7.2.6) are more precise than the error bounds given by (7.2.7).

(b) Condition $\bar{U}(x^*, R_2) \subseteq D$ can be replaced by $U(x^*, \frac{1}{K_0})$. In this case condition $(H_2)'$ holds for all $x, y \in U(x^*, \frac{1}{K_0})$.

7.3 Numerical Examples

Example 7.5 Let $D = U(1, 1)$, $x^* = \sqrt[3]{2}$ and define function F on D by

$$F(x) = x^3 - 2. \tag{7.3.1}$$

Now we are going to consider such as initial point which previous conditions cannot be satisfied but our new criteria are satisfied, that is, the improvement that we get with our new weaker criteria.

We get that

$$\eta = 0.15023\ldots$$
$$L = 4.35,$$
$$L_0 = 3.35,$$
$$\frac{1}{L_0} = 0.298507\ldots$$

and

$$L_1 = 2.94701\ldots.$$

Using this values we obtain that condition (7.1.7) is not satisfied

$$h_3 = 1.0303\ldots < 1,$$

but condition (7.1.10) is satisfied:

$$h_4 = 0.996201\ldots < 1,$$

so we can ensure the convergence of the Newton's method by Theorem 7.1.

Example 7.6 Let $X = Y = \mathcal{C}[0,1]$, the space of continuous functions defined in $[0,1]$ equipped with the max-norm. Let $\Omega = \{x \in \mathcal{C}[0,1];\ \|x\| \leq R\}$, such that $R > 0$ and F defined on Ω and given by

$$F(x)(s) = x(s) - f(s) - \lambda \int_0^1 G(s,t)x(t)^3\,dt, \quad x \in \mathcal{C}[0,1],\ s \in [0,1],$$

where $f \in \mathcal{C}[0,1]$ is a given function, λ is a real constant and the kernel G is the Green function

$$G(s,t) = \begin{cases} (1-s)t, & t \leq s, \\ s(1-t), & s \leq t. \end{cases}$$

In this case, for each $x \in \Omega$, $F'(x)$ is a linear operator defined on Ω by the following expression:

$$[F'(x)(v)](s) = v(s) - 3\lambda \int_0^1 G(s,t)x(t)^2 v(t)\,dt, \quad v \in \mathcal{C}[0,1],\ s \in [0,1].$$

If we choose $x_0(s) = f(s) = 1$, it follows $\|I - F'(x_0)\| \leq 3|\lambda|/8$. Thus, if $|\lambda| < 8/3$, $F'(x_0)^{-1}$ is defined and

$$\|F'(x_0)^{-1}\| \leq \frac{8}{8 - 3|\lambda|}.$$

Moreover,

$$\|F(x_0)\| \leq \frac{|\lambda|}{8},$$

$$\eta = \|F'(x_0)^{-1}F(x_0)\| \leq \frac{|\lambda|}{8-3|\lambda|}.$$

On the other hand, for $x, y \in \Omega$ we have

$$\|F'(x) - F'(y)\| \leq \|x-y\|\frac{1+3|\lambda|(\|x+y\|)}{8} \leq \|x-y\|\frac{1+6R|\lambda|}{8}.$$

and

$$\|F'(x) - F'(1)\| \leq \|x-1\|\frac{1+3|\lambda|(\|x\|+1)}{8} \leq \|x-1\|\frac{1+3(1+R)|\lambda|}{8}.$$

Choosing $\lambda = 1.175$ and $R = 2$, we have

$$\begin{aligned} \eta &= 0.26257\ldots, \\ L &= 2.76875\ldots, \\ L_0 &= 1.8875\ldots, \\ \frac{1}{L_0} &= 0.529801\ldots, \\ L_1 &= 1.47314\ldots. \end{aligned}$$

Using this values we obtain that condition (7.1.7) is not satisfied

$$h_3 = 1.02688\ldots < 1,$$

but condition (7.1.10) is satisfied:

$$h_4 = 0.972198\ldots < 1,$$

so we can ensure the convergence of the Newton's method by Theorem 7.1.

Example 7.7 Let $X = Y = \mathbb{R}^3$, $D = \overline{U}(0,1)$. Define F on D for $v = (x, y, z)^T$ by

$$F(v) = \left(e^x - 1, \frac{e-1}{2}y^2 + y, z\right)^T. \tag{7.3.2}$$

Then, the Fréchet-derivative is given by

$$F'(v) = \begin{bmatrix} e^x & 0 & 0 \\ 0 & (e-1)y+1 & 0 \\ 0 & 0 & 1 \end{bmatrix}.$$

Notice that $x^* = (0, 0, 0)^T$, $F'^*) = F'^*)^{-1} = diag\{1, 1, 1\}$, $K_0 = e - 1 < K = e$ and $K_1 = 1.78957397\ldots$.

With this values we obtain that our new radius is larger than the old ones as:

$$R_0 = 0.245253\ldots,$$

$$R_1 = 0.324947\ldots$$

and

$$R_0 = 0.382692\ldots.$$

References

1. S. Amat, S. Busquier, J.M. Gutiérrez, Geometric constructions of iterative functions to solve nonlinear equations. J. Comput. Appl. Math. **157**, 197–205 (2003)
2. S. Amat, S. Busquier, Third-order iterative methods under Kantorovich conditions. J. Math. Anal. Appl. **336**, 243–261 (2007)
3. S. Amat, S. Busquier, M. Negra, Adaptive approximation of nonlinear operators. Numer. Funct. Anal. Optim. **25**, 397–405 (2004)
4. I.K. Argyros, On the Newton–Kantorovich hypothesis for solving equations. J. Comput. Math. **169**, 315–332 (2004)
5. I.K. Argyros, A semilocal convergence analysis for directional Newton methods. Math. Comput. **80**, 327–343 (2011)
6. I.K. Argyros, D. González, Extending the applicability of Newton's method for k-Fréchet differentiable operators in Banach spaces. Appl. Math. Comput. **234**, 167–178 (2014)
7. I.K. Argyros, S. Hilout, Weaker conditions for the convergence of Newton's method. J. Complex., AMS **28**, 364–387 (2012)
8. I.K. Argyros, S. Hilout, *Numerical methods in Nonlinear Analysis* (World Scientific Publ. Comp, New Jersey, 2013)
9. I.K. Argyros, S. Hilout, On an improved convergence analysis of Newton's method. Appl. Math. Comput. **225**, 372–386 (2013)
10. I.K. Argyros, Á.A. Magreñán, *Extending the applicability of the local and semilocal convergence of Newton's method* (2015), submitted for publication
11. J.A. Ezquerro, M.A. Hernández, *How to improve the domain of parameters for Newton's method*, to appear in Appl. Math. Lett
12. J.M. Gutiérrez, Á.A. Magreñán, N. Romero, On the semilocal convergence of Newton–Kantorovich method under center-Lipschitz conditions. Appl. Math. Comput. **221**, 79–88 (2013)
13. L.V. Kantorovich, G.P. Akilov, *Functional Analysis* (Pergamon Press, Oxford, 1982)
14. W.C. Rheinboldt, An adaptive continuation process for solving systems of nonlinear equations. Pol. Acad. Sci., Banach Ctr. Publ. **3**, 129–142 (1978)
15. J.F. Traub, *Iterative Methods for the Solution of Equations* (Prentice- Hall Series in Automatic Computation, Englewood Cliffs, 1964)

Chapter 8
The Left Multidimensional Riemann–Liouville Fractional Integral

Here we study some important properties of left multidimensional Riemann–Liouville fractional integral operator, such as of continuity and boundedness.

8.1 Motivation

From [1], p. 388 we have:

Theorem 8.1 *Let $r > 0$, $F \in L_{\infty}(a, b)$, and*

$$G(s) = \int_{a}^{s} (s-t)^{r-1} F(t)\, dt,$$

all $s \in [a, b]$. Then $G \in AC([a, b])$ (absolutely continuous functions) for $r \geq 1$, and $G \in C([a, b])$, only for $r \in (0, 1)$.

8.2 Main Results

We give:

Theorem 8.2 *Let $f \in L_{\infty}([a, b] \times [c, d])$, $\alpha_1, \alpha_2 > 0$. Consider the function*

$$F(x_1, x_2) = \int_{a_1}^{x_1} \int_{a_2}^{x_2} (x_1 - t_1)^{\alpha_1 - 1} (x_2 - t_2)^{\alpha_2 - 1} f(t_1, t_2)\, dt_1 dt_2, \qquad (8.2.1)$$

where $a_1, x_1 \in [a, b]$, $a_2, x_2 \in [c, d]$: $a_1 \leq x_1$, $a_2 \leq x_2$.

Then F is continuous on $[a_1, b] \times [a_2, d]$.

G.A. Anastassiou and I.K. Argyros, *Intelligent Numerical Methods II: Applications to Multivariate Fractional Calculus*, Studies in Computational Intelligence 649, DOI 10.1007/978-3-319-33606-0_8

Proof (I) Let $a_1, b_1, b_1^* \in [a, b]$ with $b_1 > b_1^* > a_1$, and $a_2, b_2, b_2^* \in [c, d]$ with $b_2 > b_2^* > a_2$.

We observe that

$$\begin{aligned}
&F(b_1, b_2) - F\left(b_1^*, b_2^*\right) \\
&= \int_{a_1}^{b_1} \int_{a_2}^{b_2} (b_1 - t_1)^{\alpha_1 - 1} (b_2 - t_2)^{\alpha_2 - 1} f(t_1, t_2)\, dt_1 dt_2 \\
&\quad - \int_{a_1}^{b_1^*} \int_{a_2}^{b_2^*} \left(b_1^* - t_1\right)^{\alpha_1 - 1} \left(b_2^* - t_2\right)^{\alpha_2 - 1} f(t_1, t_2)\, dt_1 dt_2 \\
&= \int_{a_1}^{b_1^*} \int_{a_2}^{b_2^*} (b_1 - t_1)^{\alpha_1 - 1} (b_2 - t_2)^{\alpha_2 - 1} f(t_1, t_2)\, dt_1 dt_2 \\
&\quad - \int_{a_1}^{b_1^*} \int_{a_2}^{b_2^*} \left(b_1^* - t_1\right)^{\alpha_1 - 1} \left(b_2^* - t_2\right)^{\alpha_2 - 1} f(t_1, t_2)\, dt_1 dt_2 \\
&\quad + \int_{b_1^*}^{b_1} \int_{a_2}^{b_2^*} (b_1 - t_1)^{\alpha_1 - 1} (b_2 - t_2)^{\alpha_2 - 1} f(t_1, t_2)\, dt_1 dt_2 \\
&\quad + \int_{a_1}^{b_1^*} \int_{b_2^*}^{b_2} (b_1 - t_1)^{\alpha_1 - 1} (b_2 - t_2)^{\alpha_2 - 1} f(t_1, t_2)\, dt_1 dt_2 \\
&\quad + \int_{b_1^*}^{b_1} \int_{b_2^*}^{b_2} (b_1 - t_1)^{\alpha_1 - 1} (b_2 - t_2)^{\alpha_2 - 1} f(t_1, t_2)\, dt_1 dt_2.
\end{aligned} \tag{8.2.2}$$

Call

$$\begin{aligned}
&I\left(b_1^*, b_2^*\right) \\
&= \int_{a_1}^{b_1^*} \int_{a_2}^{b_2^*} \left|(b_1 - t_1)^{\alpha_1 - 1} (b_2 - t_2)^{\alpha_2 - 1} - \left(b_1^* - t_1\right)^{\alpha_1 - 1} \left(b_2^* - t_2\right)^{\alpha_2 - 1}\right| dt_1 dt_2.
\end{aligned} \tag{8.2.3}$$

Thus

$$\begin{aligned}
&\left|F(b_1, b_2) - F\left(b_1^*, b_2^*\right)\right| \\
&\leq \left\{ I\left(b_1^*, b_2^*\right) + \frac{\left(b_1 - b_1^*\right)^{\alpha_1}}{\alpha_1} \left[\frac{(b_2 - a_2)^{\alpha_2} - \left(b_2 - b_2^*\right)^{\alpha_2}}{\alpha_2}\right] \right. \\
&\left. + \left(\frac{(b_1 - a_1)^{\alpha_1} - \left(b_1 - b_1^*\right)^{\alpha_1}}{\alpha_1}\right) \frac{\left(b_2 - b_2^*\right)^{\alpha_2}}{\alpha_2} + \frac{\left(b_1 - b_1^*\right)^{\alpha_1}}{\alpha_1} \frac{\left(b_2 - b_2^*\right)^{\alpha_2}}{\alpha_2} \right\} \|f\|_\infty .
\end{aligned} \tag{8.2.4}$$

Hence, by (8.2.4), it holds

$$\delta := \lim_{\substack{(b_1^*,b_2^*)\to(b_1,b_2) \\ \text{or} \\ (b_1,b_2)\to(b_1^*,b_2^*)}} \left|F(b_1,b_2) - F(b_1^*,b_2^*)\right| \leq (\lim_{\substack{(b_1^*,b_2^*)\to(b_1,b_2) \\ \text{or} \\ (b_1,b_2)\to(b_1^*,b_2^*)}} I(b_1^*,b_2^*)) \|f\|_\infty =: \rho. \tag{8.2.5}$$

If $\alpha_1 = \alpha_2 = 1$, then $\rho = 0$, proving $\delta = 0$.

If $\alpha_1 = 1$, $\alpha_2 > 0$ we get

$$I(b_1^*,b_2^*) = (b_1^* - a_1)\left(\int_{a_2}^{b_2^*}\left|(b_2 - t_2)^{\alpha_2-1} - (b_2^* - t_2)^{\alpha_2-1}\right| dt_2\right). \tag{8.2.6}$$

Assume $\alpha_2 > 1$, then $\alpha_2 - 1 > 0$. Hence by $b_2 > b_2^*$, then $b_2 - t_2 > b_2^* - t_2 \geq 0$, and $(b_2 - t_2)^{\alpha_2-1} > (b_2^* - t_2)^{\alpha_2-1}$ and $(b_2 - t_2)^{\alpha_2-1} - (b_2^* - t_2)^{\alpha_2-1} > 0$.

That is

$$I(b_1^*,b_2^*) = (b_1^* - a_1)\left[\frac{(b_2 - t_2)^{\alpha_2}}{\alpha_2}\Bigg|_{b_2^*}^{a_2} - \frac{(b_2^* - a_2)^{\alpha_2}}{\alpha_2}\right]$$
$$= (b_1^* - a_1)\left[\frac{(b_2 - a_2)^{\alpha_2} - (b_2 - b_2^*)^{\alpha_2} - (b_2^* - a_2)^{\alpha_2}}{\alpha_2}\right]. \tag{8.2.7}$$

Clearly, then

$$\lim_{\substack{b_2^*\to b_2 \\ \text{or} \\ b_2\to b_2^*}} I(b_1^*,b_2^*) = 0. \tag{8.2.8}$$

Similarly and symmetrically, we obtain that

$$\lim_{\substack{b_1^*\to b_1 \\ \text{or} \\ b_1\to b_1^*}} I(b_1^*,b_2^*) = 0, \tag{8.2.9}$$

for the case of $\alpha_2 = 1$, $\alpha_1 > 1$.

If $\alpha_1 = 1$, and $0 < \alpha_2 < 1$, then $\alpha_2 - 1 < 0$. Hence

$$I(b_1^*,b_2^*) = (b_1^* - a_1)\left(\int_{a_2}^{b_2^*}\left((b_2^* - t_2)^{\alpha_2-1} - (b_2 - t_2)^{\alpha_2-1}\right) dt_2\right)$$
$$= (b_1^* - a_1)\left[\frac{(b_2^* - a_2)^{\alpha_2} - (b_2 - a_2)^{\alpha_2} + (b_2 - b_2^*)^{\alpha_2}}{\alpha_2}\right]. \tag{8.2.10}$$

Clearly, then

$$\lim_{\substack{b_2^* \to b_2 \\ \text{or} \\ b_2 \to b_2^*}} I\left(b_1^*, b_2^*\right) = 0. \tag{8.2.11}$$

Similarly and symmetrically, we derive that

$$\lim_{\substack{b_1^* \to b_1 \\ \text{or} \\ b_1 \to b_1^*}} I\left(b_1^*, b_2^*\right) = 0, \tag{8.2.12}$$

for the case of $\alpha_2 = 1, 0 < \alpha_1 < 1$.

Case now of $\alpha_1, \alpha_2 > 1$, then

$$\begin{aligned} & I\left(b_1^*, b_2^*\right) \\ &= \int_{a_1}^{b_1^*} \int_{a_2}^{b_2^*} \left[(b_1 - t_1)^{\alpha_1 - 1} (b_2 - t_2)^{\alpha_2 - 1} - \left(b_1^* - t_1\right)^{\alpha_1 - 1} \left(b_2^* - t_2\right)^{\alpha_2 - 1}\right] dt_1 dt_2 \\ &= \left(\frac{(b_1 - a_1)^{\alpha_1} - \left(b_1 - b_1^*\right)^{\alpha_1}}{\alpha_1}\right)\left(\frac{(b_2 - a_2)^{\alpha_2} - \left(b_2 - b_2^*\right)^{\alpha_2}}{\alpha_2}\right) \\ &\quad - \frac{\left(b_1^* - a_1\right)^{\alpha_1}}{\alpha_1} \frac{\left(b_2^* - a_2\right)^{\alpha_2}}{\alpha_2}. \end{aligned} \tag{8.2.13}$$

That is

$$\lim_{\substack{(b_1, b_2) \to \left(b_1^*, b_2^*\right) \\ \text{or} \\ \left(b_1^*, b_2^*\right) \to (b_1, b_2)}} I\left(b_1^*, b_2^*\right) = 0. \tag{8.2.14}$$

Case now of $0 < \alpha_1, \alpha_2 < 1$, then

$$\begin{aligned} & I\left(b_1^*, b_2^*\right) \\ &= \int_{a_1}^{b_1^*} \int_{a_2}^{b_2^*} \left[\left(b_1^* - t_1\right)^{\alpha_1 - 1} \left(b_2^* - t_2\right)^{\alpha_2 - 1} - (b_1 - t_1)^{\alpha_1 - 1} (b_2 - t_2)^{\alpha_2 - 1}\right] dt_1 dt_2 \\ &= \frac{\left(b_1^* - a_1\right)^{\alpha_1}}{\alpha_1} \frac{\left(b_2^* - a_2\right)^{\alpha_2}}{\alpha_2} \\ &\quad - \left(\frac{(b_1 - a_1)^{\alpha_1} - \left(b_1 - b_1^*\right)^{\alpha_1}}{\alpha_1}\right)\left(\frac{(b_2 - a_2)^{\alpha_2} - \left(b_2 - b_2^*\right)^{\alpha_2}}{\alpha_2}\right). \end{aligned} \tag{8.2.15}$$

That is again, when $0 < \alpha_1, \alpha_2 < 1$,

$$\lim_{\substack{(b_1,b_2)\to(b_1^*,b_2^*)\\ \text{or}\\ (b_1^*,b_2^*)\to(b_1,b_2)}} I\left(b_1^*, b_2^*\right) = 0. \tag{8.2.16}$$

Next we treat the case of $\alpha_1 > 1, 0 < \alpha_2 < 1$.

We observe that

$$\begin{aligned} I\left(b_1^*, b_2^*\right) &\leq I^*\left(b_1^*, b_2^*\right) \\ &:= \int_{a_1}^{b_1^*}\int_{a_2}^{b_2^*} (b_1 - t_1)^{\alpha_1 - 1}\left|(b_2 - t_2)^{\alpha_2 - 1} - \left(b_2^* - t_2\right)^{\alpha_2 - 1}\right| dt_1 dt_2 \\ &+ \int_{a_1}^{b_1^*}\int_{a_2}^{b_2^*} \left(b_2^* - t_2\right)^{\alpha_2 - 1}\left|(b_1 - t_1)^{\alpha_1 - 1} - \left(b_1^* - t_1\right)^{\alpha_1 - 1}\right| dt_1 dt_2. \end{aligned} \tag{8.2.17}$$

Therefore it holds

$$\begin{aligned} I^*\left(b_1^*, b_2^*\right) &= \int_{a_1}^{b_1^*}\int_{a_2}^{b_2^*} (b_1 - t_1)^{\alpha_1 - 1}\left(\left(b_2^* - t_2\right)^{\alpha_2 - 1} - (b_2 - t_2)^{\alpha_2 - 1}\right) dt_1 dt_2 \qquad (8.2.18) \\ &+ \int_{a_1}^{b_1^*}\int_{a_2}^{b_2^*} \left(b_2^* - t_2\right)^{\alpha_2 - 1}\left((b_1 - t_1)^{\alpha_1 - 1} - \left(b_1^* - t_1\right)^{\alpha_1 - 1}\right) dt_1 dt_2 \\ &= \left(\frac{(b_1 - a_1)^{\alpha_1} - \left(b_1 - b_1^*\right)^{\alpha_1}}{\alpha_1}\right)\left[\frac{\left(b_2^* - a_2\right)^{\alpha_2} - (b_2 - a_2)^{\alpha_2} + \left(b_2 - b_2^*\right)^{\alpha_2}}{\alpha_2}\right] \\ &+ \frac{\left(b_2^* - a_2\right)^{\alpha_2}}{\alpha_2}\left[\frac{(b_1 - a_1)^{\alpha_1} - \left(b_1 - b_1^*\right)^{\alpha_1} - \left(b_1^* - a_1\right)^{\alpha_1}}{\alpha_1}\right]. \qquad (8.2.19) \end{aligned}$$

So, in case of $\alpha_1 > 1, 0 < \alpha_2 < 1$, we proved that

$$\lim_{\substack{(b_1,b_2)\to(b_1^*,b_2^*)\\ \text{or}\\ (b_1^*,b_2^*)\to(b_1,b_2)}} I\left(b_1^*, b_2^*\right) = 0. \tag{8.2.20}$$

Finally, we prove the case of $\alpha_2 > 1$ and $0 < \alpha_1 < 1$. We have that

$$\begin{aligned} I^*\left(b_1^*, b_2^*\right) &\overset{(8.2.17)}{=} \int_{a_1}^{b_1^*}\int_{a_2}^{b_2^*} (b_1 - t_1)^{\alpha_1 - 1}\left[(b_2 - t_2)^{\alpha_2 - 1} - \left(b_2^* - t_2\right)^{\alpha_2 - 1}\right] dt_1 dt_2 \\ &+ \int_{a_1}^{b_1^*}\int_{a_2}^{b_2^*} \left(b_2^* - t_2\right)^{\alpha_2 - 1}\left(\left(b_1^* - t_1\right)^{\alpha_1 - 1} - (b_1 - t_1)^{\alpha_1 - 1}\right) dt_1 dt_2 \end{aligned} \tag{8.2.21}$$

$$= \left(\frac{(b_1 - a_1)^{\alpha_1} - (b_1 - b_1^*)^{\alpha_1}}{\alpha_1} \right) \left[\frac{(b_2 - a_2)^{\alpha_2} - (b_2 - b_2^*)^{\alpha_2} - (b_2^* - a_2)^{\alpha_2}}{\alpha_2} \right]$$

$$+ \frac{(b_2^* - a_2)^{\alpha_2}}{\alpha_2} \left[\frac{(b_1^* - a_1)^{\alpha_1} - (b_1 - a_1)^{\alpha_1} + (b_1 - b_1^*)^{\alpha_1}}{\alpha_1} \right]. \quad (8.2.22)$$

Hence again it holds

$$\lim_{\substack{(b_1, b_2) \to (b_1^*, b_2^*) \\ \text{or} \\ (b_1^*, b_2^*) \to (b_1, b_2)}} I\left(b_1^*, b_2^*\right) = 0. \quad (8.2.23)$$

We proved $\rho = 0$, and $\delta = 0$ in all cases of this section.

The case of $b_1^* > b_1$ and $b_2^* > b_2$, as symmetric to $b_1 > b_1^*$ and $b_2 > b_2^*$ we treated, it is omitted, a totally similar treatment.

(II) The remaining cases are: let $a_1, b_1, b_1^* \in [a, b]$; $a_2, b_2, b_2^* \in [c, d]$, we can have

(II_1) $b_1 > b_1^*$ and $b_2 < b_2^*$,

or

(II_2) $b_1 < b_1^*$ and $b_2 > b_2^*$.

Notice that (II_1) and (II_2) cases are symmetric, and treated the same way. As such we treat only the case (II_1).

We observe again that

$$F(b_1, b_2) - F\left(b_1^*, b_2^*\right)$$
$$= \int_{a_1}^{b_1} \int_{a_2}^{b_2} (b_1 - t_1)^{\alpha_1 - 1} (b_2 - t_2)^{\alpha_2 - 1} f(t_1, t_2)\, dt_1 dt_2$$
$$- \int_{a_1}^{b_1^*} \int_{a_2}^{b_2^*} \left(b_1^* - t_1\right)^{\alpha_1 - 1} \left(b_2^* - t_2\right)^{\alpha_2 - 1} f(t_1, t_2)\, dt_1 dt_2 \quad (8.2.24)$$
$$= \int_{a_1}^{b_1^*} \int_{a_2}^{b_2} (b_1 - t_1)^{\alpha_1 - 1} (b_2 - t_2)^{\alpha_2 - 1} f(t_1, t_2)\, dt_1 dt_2$$
$$+ \int_{b_1^*}^{b_1} \int_{a_2}^{b_2} (b_1 - t_1)^{\alpha_1 - 1} (b_2 - t_2)^{\alpha_2 - 1} f(t_1, t_2)\, dt_1 dt_2$$
$$- \int_{a_1}^{b_1^*} \int_{a_2}^{b_2} \left(b_1^* - t_1\right)^{\alpha_1 - 1} \left(b_2^* - t_2\right)^{\alpha_2 - 1} f(t_1, t_2)\, dt_1 dt_2$$
$$- \int_{a_1}^{b_1^*} \int_{b_2}^{b_2^*} \left(b_1^* - t_1\right)^{\alpha_1 - 1} \left(b_2^* - t_2\right)^{\alpha_2 - 1} f(t_1, t_2)\, dt_1 dt_2$$
$$= \int_{a_1}^{b_1^*} \int_{a_2}^{b_2} \left((b_1 - t_1)^{\alpha_1 - 1} (b_2 - t_2)^{\alpha_2 - 1} - \left(b_1^* - t_1\right)^{\alpha_1 - 1} \left(b_2^* - t_2\right)^{\alpha_2 - 1}\right) f(t_1, t_2)\, dt_1 dt_2$$
$$+ \int_{b_1^*}^{b_1} \int_{a_2}^{b_2} (b_1 - t_1)^{\alpha_1 - 1} (b_2 - t_2)^{\alpha_2 - 1} f(t_1, t_2)\, dt_1 dt_2$$
$$- \int_{a_1}^{b_1^*} \int_{b_2}^{b_2^*} \left(b_1^* - t_1\right)^{\alpha_1 - 1} \left(b_2^* - t_2\right)^{\alpha_2 - 1} f(t_1, t_2)\, dt_1 dt_2. \quad (8.2.25)$$

We call

$$I\left(b_1^*, b_2\right) := \int_{a_1}^{b_1^*} \int_{a_2}^{b_2} \left| (b_1 - t_1)^{\alpha_1 - 1} (b_2 - t_2)^{\alpha_2 - 1} - (b_1^* - t_1)^{\alpha_1 - 1} (b_2^* - t_2)^{\alpha_2 - 1} \right| dt_1 dt_2. \tag{8.2.26}$$

Hence, we have

$$\left|F(b_1, b_2) - F\left(b_1^*, b_2^*\right)\right| \leq \left\{ I\left(b_1^*, b_2\right) + \frac{\left(b_1 - b_1^*\right)^{\alpha_1}}{\alpha_1} \frac{(b_2 - a_2)^{\alpha_2}}{\alpha_2} + \frac{\left(b_1^* - a_1\right)^{\alpha_1}}{\alpha_1} \frac{\left(b_2^* - b_2\right)^{\alpha_2}}{\alpha_2} \right\} \|f\|_\infty . \tag{8.2.27}$$

Therefore it holds

$$\delta := \lim_{\substack{|b_1 - b_1^*| \to 0, \\ |b_2 - b_2^*| \to 0}} \left|F(b_1, b_2) - F\left(b_1^*, b_2^*\right)\right| \leq \left(\lim_{\substack{|b_1 - b_1^*| \to 0, \\ |b_2 - b_2^*| \to 0}} I\left(b_1^*, b_2\right) \right) \|f\|_\infty =: \theta. \tag{8.2.28}$$

We will prove that $\theta = 0$, hence $\delta = 0$, in all possible cases.

If $\alpha_1 = \alpha_2 = 1$, then $I\left(b_1^*, b_2\right) = 0$, hence $\theta = 0$.

If $\alpha_1 = 1$, $\alpha_2 > 0$ we get

$$I\left(b_1^*, b_2\right) = \left(b_1^* - a_1\right) \left(\int_{a_2}^{b_2} \left| (b_2 - t_2)^{\alpha_2 - 1} - \left(b_2^* - t_2\right)^{\alpha_2 - 1} \right| dt_2 \right). \tag{8.2.29}$$

Assume $\alpha_2 > 1$, then $\alpha_2 - 1 > 0$. Hence

$$\begin{aligned} I\left(b_1^*, b_2\right) &= \left(b_1^* - a_1\right) \left(\int_{a_2}^{b_2} \left(\left(b_2^* - t_2\right)^{\alpha_2 - 1} - (b_2 - t_2)^{\alpha_2 - 1} \right) dt_2 \right) \\ &= \left(b_1^* - a_1\right) \left[\frac{\left(b_2^* - t_2\right)^{\alpha_2} \Big|_{b_2}^{a_2} - (b_2 - a_2)^{\alpha_2}}{\alpha_2} \right] \\ &= \left(b_1^* - a_1\right) \left[\frac{\left(b_2^* - a_2\right)^{\alpha_2} - \left(b_2^* - b_2\right)^{\alpha_2} - (b_2 - a_2)^{\alpha_2}}{\alpha_2} \right]. \end{aligned} \tag{8.2.30}$$

Clearly, then

$$\lim_{\substack{|b_1 - b_1^*| \to 0, \\ |b_2 - b_2^*| \to 0}} I\left(b_1^*, b_2\right) = 0, \tag{8.2.31}$$

hence $\theta = 0$.

Let the case now of $\alpha_2 = 1$, $\alpha_1 > 1$. Then

$$I\left(b_1^*, b_2\right) = (b_2 - a_2)\left(\int_{a_1}^{b_1^*} \left|(b_1 - t_1)^{\alpha_1 - 1} - \left(b_1^* - t_1\right)^{\alpha_1 - 1}\right| dt_1\right)$$
$$= (b_2 - a_2)\left[\frac{(b_1 - a_1)^{\alpha_1} - \left(b_1 - b_1^*\right)^{\alpha_1} - \left(b_1^* - a_1\right)^{\alpha_1}}{\alpha_1}\right]. \tag{8.2.32}$$

Then $\theta = 0$.

If $\alpha_1 = 1$, and $0 < \alpha_2 < 1$, then $\alpha_2 - 1 < 0$. Hence

$$I\left(b_1^*, b_2\right) = \left(b_1^* - a_1\right)\int_{a_2}^{b_2} \left|(b_2 - t_2)^{\alpha_2 - 1} - \left(b_2^* - t_2\right)^{\alpha_2 - 1}\right| dt_2$$
$$= \left(b_1^* - a_1\right)\int_{a_2}^{b_2} \left((b_2 - t_2)^{\alpha_2 - 1} - \left(b_2^* - t_2\right)^{\alpha_2 - 1}\right) dt_2$$
$$= \left(b_1^* - a_1\right)\left[\frac{(b_2 - a_2)^{\alpha_2} - \left(b_2^* - a_2\right)^{\alpha_2} + \left(b_2^* - b_2\right)^{\alpha_2}}{\alpha_2}\right]. \tag{8.2.33}$$

Hence $\theta = 0$.

Let now $\alpha_2 = 1$, $0 < \alpha_1 < 1$. Then

$$I\left(b_1^*, b_2\right) = (b_2 - a_2)\int_{a_1}^{b_1^*} \left|(b_1 - t_1)^{\alpha_1 - 1} - \left(b_1^* - t_1\right)^{\alpha_1 - 1}\right| dt_1$$
$$= (b_2 - a_2)\int_{a_1}^{b_1^*} \left(\left(b_1^* - t_1\right)^{\alpha_1 - 1} - (b_1 - t_1)^{\alpha_1 - 1}\right) dt_1$$
$$= (b_2 - a_2)\left[\frac{\left(b_1^* - a_1\right)^{\alpha_1} - (b_1 - a_1)^{\alpha_1} + \left(b_1 - b_1^*\right)^{\alpha_1}}{\alpha_1}\right]. \tag{8.2.34}$$

Hence $\theta = 0$.

We observe that:

$$I\left(b_1^*, b_2\right)$$
$$\leq \int_{a_1}^{b_1^*}\int_{a_2}^{b_2} \left|(b_1 - t_1)^{\alpha_1 - 1}(b_2 - t_2)^{\alpha_2 - 1} - (b_1 - t_1)^{\alpha_1 - 1}\left(b_2^* - t_2\right)^{\alpha_2 - 1}\right| dt_1 dt_2$$
$$+ \int_{a_1}^{b_1^*}\int_{a_2}^{b_2} \left|(b_1 - t_1)^{\alpha_1 - 1}\left(b_2^* - t_2\right)^{\alpha_2 - 1} - \left(b_1^* - t_1\right)^{\alpha_1 - 1}\left(b_2^* - t_2\right)^{\alpha_2 - 1}\right| dt_1 dt_2 \tag{8.2.35}$$
$$=: J\left(b_1^*, b_2\right),$$

i.e.

$$I\left(b_1^*, b_2\right) \leq J\left(b_1^*, b_2\right).$$

Hence it holds

$$J\left(b_1^*, b_2\right) = \int_{a_1}^{b_1^*}\int_{a_2}^{b_2} (b_1 - t_1)^{\alpha_1 - 1}\left|(b_2 - t_2)^{\alpha_2 - 1} - \left(b_2^* - t_2\right)^{\alpha_2 - 1}\right| dt_1 dt_2$$

$$+ \int_{a_1}^{b_1^*}\int_{a_2}^{b_2} \left(b_2^* - t_2\right)^{\alpha_2 - 1}\left|(b_1 - t_1)^{\alpha_1 - 1} - \left(b_1^* - t_1\right)^{\alpha_1 - 1}\right| dt_1 dt_2. \quad (8.2.36)$$

Case of $\alpha_1, \alpha_2 > 1$. Then

$$J\left(b_1^*, b_2\right) = \int_{a_1}^{b_1^*}\int_{a_2}^{b_2} (b_1 - t_1)^{\alpha_1 - 1}\left(\left(b_2^* - t_2\right)^{\alpha_2 - 1} - (b_2 - t_2)^{\alpha_2 - 1}\right) dt_1 dt_2$$

$$+ \int_{a_1}^{b_1^*}\int_{a_2}^{b_2} \left(b_2^* - t_2\right)^{\alpha_2 - 1}\left((b_1 - t_1)^{\alpha_1 - 1} - \left(b_1^* - t_1\right)^{\alpha_1 - 1}\right) dt_1 dt_2$$

$$= \left(\frac{(b_1 - a_1)^{\alpha_1} - \left(b_1 - b_1^*\right)^{\alpha_1}}{\alpha_1}\right)\left[\left(\frac{\left(b_2^* - a_2\right)^{\alpha_2} - \left(b_2^* - b_2\right)^{\alpha_2}}{\alpha_2}\right) - \frac{(b_2 - a_2)^{\alpha_2}}{\alpha_2}\right]$$

$$+ \left(\frac{\left(b_2^* - a_2\right)^{\alpha_2} - \left(b_2^* - b_2\right)^{\alpha_2}}{\alpha_2}\right)\left[\left(\frac{(b_1 - a_1)^{\alpha_1} - \left(b_1 - b_1^*\right)^{\alpha_1}}{\alpha_1}\right) - \frac{\left(b_1^* - a_1\right)^{\alpha_1}}{\alpha_1}\right]. \quad (8.2.37)$$

So that $\theta = 0$.

Case of $0 < \alpha_1, \alpha_2 < 1$, then

$$J\left(b_1^*, b_2\right) = \int_{a_1}^{b_1^*}\int_{a_2}^{b_2} (b_1 - t_1)^{\alpha_1 - 1}\left((b_2 - t_2)^{\alpha_2 - 1} - \left(b_2^* - t_2\right)^{\alpha_2 - 1}\right) dt_1 dt_2$$

$$+ \int_{a_1}^{b_1^*}\int_{a_2}^{b_2} \left(b_2^* - t_2\right)^{\alpha_2 - 1}\left(\left(b_1^* - t_1\right)^{\alpha_1 - 1} - (b_1 - t_1)^{\alpha_1 - 1}\right) dt_1 dt_2$$

$$= \left(\frac{(b_1 - a_1)^{\alpha_1} - \left(b_1 - b_1^*\right)^{\alpha_1}}{\alpha_1}\right)\left[\frac{(b_2 - a_2)^{\alpha_2}}{\alpha_2} - \left(\frac{\left(b_2^* - a_2\right)^{\alpha_2} - \left(b_2^* - b_2\right)^{\alpha_2}}{\alpha_2}\right)\right]$$

$$+ \left(\frac{\left(b_2^* - a_2\right)^{\alpha_2} - \left(b_2^* - b_2\right)^{\alpha_2}}{\alpha_2}\right)\left[\frac{\left(b_1^* - a_1\right)^{\alpha_1}}{\alpha_1} - \left(\frac{(b_1 - a_1)^{\alpha_1} - \left(b_1 - b_1^*\right)^{\alpha_1}}{\alpha_1}\right)\right]. \quad (8.2.38)$$

One more time $\theta = 0$.

Next case of $\alpha_1 > 1, 0 < \alpha_2 < 1$. We observe that

$$J\left(b_1^*, b_2\right) = \int_{a_1}^{b_1^*} \int_{a_2}^{b_2} (b_1 - t_1)^{\alpha_1 - 1} \left((b_2 - t_2)^{\alpha_2 - 1} - \left(b_2^* - t_2\right)^{\alpha_2 - 1}\right) dt_1 dt_2 \tag{8.2.39}$$

$$+ \int_{a_1}^{b_1^*} \int_{a_2}^{b_2} \left(b_2^* - t_2\right)^{\alpha_2 - 1} \left((b_1 - t_1)^{\alpha_1 - 1} - \left(b_1^* - t_1\right)^{\alpha_1 - 1}\right) dt_1 dt_2$$

$$= \left(\frac{(b_1 - a_1)^{\alpha_1} - \left(b_1 - b_1^*\right)^{\alpha_1}}{\alpha_1}\right) \left[\frac{(b_2 - a_2)^{\alpha_2}}{\alpha_2} - \left(\frac{\left(b_2^* - a_2\right)^{\alpha_2} - \left(b_2^* - b_2\right)^{\alpha_2}}{\alpha_2}\right)\right]$$

$$+ \left(\frac{\left(b_2^* - a_2\right)^{\alpha_2} - \left(b_2^* - b_2\right)^{\alpha_2}}{\alpha_2}\right) \left[\left(\frac{(b_1 - a_1)^{\alpha_1} - \left(b_1 - b_1^*\right)^{\alpha_1}}{\alpha_1}\right) - \frac{\left(b_1^* - a_1\right)^{\alpha_1}}{\alpha_1}\right]. \tag{8.2.40}$$

Hence $\theta = 0$.

Finally, we prove the case of $\alpha_2 > 1$ and $0 < \alpha_1 < 1$. In that case it holds

$$J\left(b_1^*, b_2\right) = \int_{a_1}^{b_1^*} \int_{a_2}^{b_2} (b_1 - t_1)^{\alpha_1 - 1} \left(\left(b_2^* - t_2\right)^{\alpha_2 - 1} - (b_2 - t_2)^{\alpha_2 - 1}\right) dt_1 dt_2 \tag{8.2.41}$$

$$+ \int_{a_1}^{b_1^*} \int_{a_2}^{b_2} \left(b_2^* - t_2\right)^{\alpha_2 - 1} \left(\left(b_1^* - t_1\right)^{\alpha_1 - 1} - (b_1 - t_1)^{\alpha_1 - 1}\right) dt_1 dt_2$$

$$= \left(\frac{(b_1 - a_1)^{\alpha_1} - \left(b_1 - b_1^*\right)^{\alpha_1}}{\alpha_1}\right) \left[-\frac{(b_2 - a_2)^{\alpha_2}}{\alpha_2} + \left(\frac{\left(b_2^* - a_2\right)^{\alpha_2} - \left(b_2^* - b_2\right)^{\alpha_2}}{\alpha_2}\right)\right]$$

$$+ \left(\frac{\left(b_2^* - a_2\right)^{\alpha_2} - \left(b_2^* - b_2\right)^{\alpha_2}}{\alpha_2}\right) \left[-\left(\frac{(b_1 - a_1)^{\alpha_1} - \left(b_1 - b_1^*\right)^{\alpha_1}}{\alpha_1}\right) + \frac{\left(b_1^* - a_1\right)^{\alpha_1}}{\alpha_1}\right]. \tag{8.2.42}$$

Hence again $\theta = 0$.

We have proved that $\delta = 0$, in all possible subcases of (II_1).

We have proved that F is a continuous function over $[a_1, b] \times [a_2, d]$.

Now we can state:

Theorem 8.3 *Let* $f \in L_\infty\left(\prod_{i=1}^k [a_i, b_i]\right)$, $\alpha_i > 0$, $i = 1, \ldots, k \in \mathbb{N}$. *Consider the function*

$$F(x_1, \ldots, x_k) = \int_{a_1^*}^{x_1} \ldots \int_{a_k^*}^{x_k} \prod_{i=1}^k (x_i - t_i)^{\alpha_i - 1} f(t_1, \ldots, t_k)\, dt_1 \ldots dt_k, \tag{8.2.43}$$

where $a_i^*, x_i \in [a_i, b_i]$, $a_i^* \le x_i$, $i = 1, \ldots, k$.

Then F *is continuous on* $\prod_{i=1}^k \left[a_i^*, b_i\right]$.

Remark 8.4 In the setting of Theorem 8.3: Consider the left multidimensional Riemann–Liouville fractional integral of order $\alpha = (\alpha_1, \ldots, \alpha_k)$:

$$\left(I_{a_+^*}^{\alpha} f\right)(x) = \frac{1}{\prod_{i=1}^k \Gamma(\alpha_i)} \int_{a_1^*}^{x_1} \dots \int_{a_k^*}^{x_k} \prod_{i=1}^k (x_i - t_i)^{\alpha_i - 1} f(t_1, \dots, t_k)\, dt_1 \dots dt_k, \tag{8.2.44}$$

where $a^* = \left(a_1^*, \dots, a_k^*\right)$, $x = (x_1, \dots, x_k)$, $a_i^* \leq x_i$, $i = 1, \dots, k$. Here Γ denotes the gamma function.

By Theorem 8.3 we get that $\left(I_{a_+^*}^{\alpha} f\right)(x)$ is a continuous function for every $x \in \prod_{i=1}^k \left[a_i^*, b_i\right]$.

We notice that

$$\begin{aligned}
\left|\left(I_{a_+^*}^{\alpha} f\right)(x)\right| &\leq \frac{1}{\prod_{i=1}^k \Gamma(\alpha_i)} \left(\int_{a_1^*}^{x_1} \dots \int_{a_k^*}^{x_k} \prod_{i=1}^k (x_i - t_i)^{\alpha_i - 1} dt_1 \dots dt_k\right) \|f\|_\infty \\
&= \frac{\|f\|_\infty}{\prod_{i=1}^k \Gamma(\alpha_i)} \prod_{i=1}^k \left(\int_{a_i^*}^{x_i} (x_i - t_i)^{\alpha_i - 1} dt_i\right) = \frac{\|f\|_\infty}{\prod_{i=1}^k \Gamma(\alpha_i)} \prod_{i=1}^k \frac{\left(x_i - a_i^*\right)^{\alpha_i}}{\alpha_i} \qquad (8.2.45) \\
&= \|f\|_\infty \left(\prod_{i=1}^k \frac{\left(x_i - a_i^*\right)^{\alpha_i}}{\Gamma(\alpha_i + 1)}\right).
\end{aligned}$$

That is

$$\left|\left(I_{a_+^*}^{\alpha} f\right)(x)\right| \leq \left(\prod_{i=1}^k \frac{\left(x_i - a_i^*\right)^{\alpha_i}}{\Gamma(\alpha_i + 1)}\right) \|f\|_\infty. \tag{8.2.46}$$

In particular we get that

$$\left(I_{a_+^*}^{\alpha} f\right)\left(a^*\right) = 0, \tag{8.2.47}$$

and

$$\left\|I_{a_+^*}^{\alpha} f\right\|_\infty \leq \left(\prod_{i=1}^k \frac{\left(b_i - a_i^*\right)^{\alpha_i}}{\Gamma(\alpha_i + 1)}\right) \|f\|_\infty. \tag{8.2.48}$$

That is $I_{a_+^*}^{\alpha} f$ is a bounded linear operator, which here is also a positive operator.

Reference

1. G. Anastassiou, *Fractional Differentiation Inequalities* (Springer, New York, 2009)

Chapter 9
The Right Multidimensional Riemann–Liouville Fractional Integral

Here we study some important properties of right multidimensional Riemann–Liouville fractional integral operator, such as of continuity and boundedness.

9.1 Motivation

From [1] we have:

Theorem 9.1 *Let* $r > 0$, $F \in L_{\infty}(a, b)$, *and*

$$G(s) = \int_{s}^{b} (t-s)^{r-1} F(t)\, dt,$$

all $s \in [a, b]$. *Then* $G \in AC([a, b])$ *(absolutely continuous functions) for* $r \geq 1$, *and* $G \in C([a, b])$, *only for* $r \in (0, 1)$.

9.2 Main Results

We give:

Theorem 9.2 *Let* $f \in L_{\infty}([a, b] \times [c, d])$, $\alpha_1, \alpha_2 > 0$. *Consider the function*

$$F(x_1, x_2) = \int_{x_1}^{b_1} \int_{x_2}^{b_2} (t_1 - x_1)^{\alpha_1 - 1} (t_2 - x_2)^{\alpha_2 - 1} f(t_1, t_2)\, dt_1 dt_2, \tag{9.2.1}$$

where $x_1, b_1 \in [a, b]$, $x_2, b_2 \in [c, d] : x_1 \leq b_1$, $x_2 \leq b_2$.

Then F *is continuous on* $[a, b_1] \times [c, b_2]$.

G.A. Anastassiou and I.K. Argyros, *Intelligent Numerical Methods II: Applications to Multivariate Fractional Calculus*, Studies in Computational Intelligence 649, DOI 10.1007/978-3-319-33606-0_9

Proof (I) Let $a_1, a_1^*, b_1 \in [a, b]$ with $a_1 < a_1^* < b_1$, and $a_2, a_2^*, b_2 \in [c, d]$ with $a_2 < a_2^* < b_2$.

We observe that

$$F(a_1, a_2) - F(a_1^*, a_2^*) =$$

$$\int_{a_1}^{b_1} \int_{a_2}^{b_2} (t_1 - a_1)^{\alpha_1 - 1} (t_2 - a_2)^{\alpha_2 - 1} f(t_1, t_2) \, dt_1 dt_2 -$$

$$\int_{a_1^*}^{b_1} \int_{a_2^*}^{b_2} (t_1 - a_1^*)^{\alpha_1 - 1} (t_2 - a_2^*)^{\alpha_2 - 1} f(t_1, t_2) \, dt_1 dt_2 = \tag{9.2.2}$$

$$\int_{a_1^*}^{b_1} \int_{a_2^*}^{b_2} (t_1 - a_1)^{\alpha_1 - 1} (t_2 - a_2)^{\alpha_2 - 1} f(t_1, t_2) \, dt_1 dt_2 +$$

$$\int_{a_1^*}^{b_1} \int_{a_2}^{a_2^*} (t_1 - a_1)^{\alpha_1 - 1} (t_2 - a_2)^{\alpha_2 - 1} f(t_1, t_2) \, dt_1 dt_2 +$$

$$\int_{a_1}^{a_1^*} \int_{a_2}^{a_2^*} (t_1 - a_1)^{\alpha_1 - 1} (t_2 - a_2)^{\alpha_2 - 1} f(t_1, t_2) \, dt_1 dt_2 +$$

$$\int_{a_1}^{a_1^*} \int_{a_2^*}^{b_2} (t_1 - a_1)^{\alpha_1 - 1} (t_2 - a_2)^{\alpha_2 - 1} f(t_1, t_2) \, dt_1 dt_2 - \tag{9.2.3}$$

$$\int_{a_1^*}^{b_1} \int_{a_2^*}^{b_2} (t_1 - a_1^*)^{\alpha_1 - 1} (t_2 - a_2^*)^{\alpha_2 - 1} f(t_1, t_2) \, dt_1 dt_2 =$$

$$\int_{a_1^*}^{b_1} \int_{a_2^*}^{b_2} \left[(t_1 - a_1)^{\alpha_1 - 1} (t_2 - a_2)^{\alpha_2 - 1} - (t_1 - a_1^*)^{\alpha_1 - 1} (t_2 - a_2^*)^{\alpha_2 - 1} \right] f(t_1, t_2) \, dt_1 dt_2$$

$$+ \int_{a_1^*}^{b_1} \int_{a_2}^{a_2^*} (t_1 - a_1)^{\alpha_1 - 1} (t_2 - a_2)^{\alpha_2 - 1} f(t_1, t_2) \, dt_1 dt_2 + \tag{9.2.4}$$

$$\int_{a_1}^{a_1^*} \int_{a_2}^{a_2^*} (t_1 - a_1)^{\alpha_1 - 1} (t_2 - a_2)^{\alpha_2 - 1} f(t_1, t_2) \, dt_1 dt_2 +$$

$$\int_{a_1}^{a_1^*} \int_{a_2^*}^{b_2} (t_1 - a_1)^{\alpha_1 - 1} (t_2 - a_2)^{\alpha_2 - 1} f(t_1, t_2) \, dt_1 dt_2.$$

Call

$$I\left(a_1^*, a_2^*\right) =$$

$$\int_{a_1^*}^{b_1} \int_{a_2^*}^{b_2} \left| (t_1 - a_1)^{\alpha_1 - 1} (t_2 - a_2)^{\alpha_2 - 1} - \left(t_1 - a_1^*\right)^{\alpha_1 - 1} \left(t_2 - a_2^*\right)^{\alpha_2 - 1} \right| dt_1 dt_2. \tag{9.2.5}$$

Thus

$$\left|F(a_1, a_2) - F\left(a_1^*, a_2^*\right)\right| \leq$$

$$\left\{ I\left(a_1^*, a_2^*\right) + \left(\frac{(b_1 - a_1)^{\alpha_1} - \left(a_1^* - a_1\right)^{\alpha_1}}{\alpha_1} \right) \frac{\left(a_2^* - a_2\right)^{\alpha_2}}{\alpha_2} + \right.$$

$$\left. \frac{\left(a_1^* - a_1\right)^{\alpha_1}}{\alpha_1} \frac{\left(a_2^* - a_2\right)^{\alpha_2}}{\alpha_2} + \frac{\left(a_1^* - a_1\right)^{\alpha_1}}{\alpha_1} \left(\frac{(b_2 - a_2)^{\alpha_2} - \left(a_2^* - a_2\right)^{\alpha_2}}{\alpha_2} \right) \right\} \|f\|_\infty .$$

Hence, by the last inequality, it holds

$$\delta := \lim_{\substack{\left(a_1^*, a_2^*\right) \to (a_1, a_2) \\ \text{or} \\ (a_1, a_2) \to \left(a_1^*, a_2^*\right)}} \left|F(a_1, a_2) - F\left(a_1^*, a_2^*\right)\right| \leq$$

$$\left(\lim_{\substack{\left(a_1^*, a_2^*\right) \to (a_1, a_2) \\ \text{or} \\ (a_1, a_2) \to \left(a_1^*, a_2^*\right)}} I\left(a_1^*, a_2^*\right) \right) \|f\|_\infty =: \rho. \tag{9.2.6}$$

If $\alpha_1 = \alpha_2 = 1$, then $\rho = 0$, proving $\delta = 0$.

If $\alpha_1 = 1$, $\alpha_2 > 0$ we get

$$I\left(a_1^*, a_2^*\right) = \left(b_1 - a_1^*\right) \int_{a_2^*}^{b_2} \left| (t_2 - a_2)^{\alpha_2 - 1} - \left(t_2 - a_2^*\right)^{\alpha_2 - 1} \right| dt_2. \tag{9.2.7}$$

Assume $\alpha_2 > 1$, then $\alpha_2 - 1 > 0$. Hence

$$I\left(a_1^*, a_2^*\right) = \left(b_1 - a_1^*\right) \int_{a_2^*}^{b_2} \left((t_2 - a_2)^{\alpha_2 - 1} - \left(t_2 - a_2^*\right)^{\alpha_2 - 1} \right) dt_2$$

$$= \left(b_1 - a_1^*\right) \left\{ \left(\frac{(b_2 - a_2)^{\alpha_2} - \left(a_2^* - a_2\right)^{\alpha_2}}{\alpha_2} \right) - \frac{\left(b_2 - a_2^*\right)^{\alpha_2}}{\alpha_2} \right\}. \tag{9.2.8}$$

Clearly, then

$$\lim_{\substack{(a_1,a_2)\to(a_1^*,a_2^*)\\ \text{or}\\ (a_1^*,a_2^*)\to(a_1,a_2)}} I\left(a_1^*,a_2^*\right)=0. \tag{9.2.9}$$

Similarly and symmetrically, we obtain that

$$\lim_{\substack{(a_1,a_2)\to(a_1^*,a_2^*)\\ \text{or}\\ (a_1^*,a_2^*)\to(a_1,a_2)}} I\left(a_1^*,a_2^*\right)=0 \tag{9.2.10}$$

for the case of $\alpha_2=1$, $\alpha_1>1$.

If $\alpha_1=1$, and $0<\alpha_2<1$, then $\alpha_2-1<0$. Hence

$$I\left(a_1^*,a_2^*\right)=\left(b_1-a_1^*\right)\int_{a_2^*}^{b_2}\left[\left(t_2-a_2^*\right)^{\alpha_2-1}-\left(t_2-a_2\right)^{\alpha_2-1}\right]dt_2=$$

$$\left(b_1-a_1^*\right)\left\{\frac{\left(b_2-a_2^*\right)^{\alpha_2}}{\alpha_2}-\left(\frac{\left(b_2-a_2\right)^{\alpha_2}-\left(a_2^*-a_2\right)^{\alpha_2}}{\alpha_2}\right)\right\}. \tag{9.2.11}$$

Clearly, then

$$\lim_{\substack{a_2^*\to a_2\\ \text{or}\\ a_2\to a_2^*}} I\left(a_1^*,a_2^*\right)=0. \tag{9.2.12}$$

Similarly and symmetrically, we derive that

$$\lim_{\substack{a_1^*\to a_1\\ \text{or}\\ a_1\to a_1^*}} I\left(a_1^*,a_2^*\right)=0, \tag{9.2.13}$$

for the case of $\alpha_2=1$, $0<\alpha_1<1$.

Case now of $\alpha_1,\alpha_2>1$, then

$$I\left(a_1^*,a_2^*\right)=$$

$$\int_{a_1^*}^{b_1}\int_{a_2^*}^{b_2}\left(\left(t_1-a_1\right)^{\alpha_1-1}\left(t_2-a_2\right)^{\alpha_2-1}-\left(t_1-a_1^*\right)^{\alpha_1-1}\left(t_2-a_2^*\right)^{\alpha_2-1}\right)dt_1dt_2$$

$$=\left(\frac{\left(b_1-a_1\right)^{\alpha_1}-\left(a_1^*-a_1\right)^{\alpha_1}}{\alpha_1}\right)\left(\frac{\left(b_2-a_2\right)^{\alpha_2}-\left(a_2^*-a_2\right)^{\alpha_2}}{\alpha_2}\right)$$

$$-\frac{\left(b_1-a_1^*\right)^{\alpha_1}}{\alpha_1}\frac{\left(b_2-a_2^*\right)^{\alpha_2}}{\alpha_2}. \tag{9.2.14}$$

That is

$$\lim_{\substack{(a_1^*,a_2^*)\to(a_1,a_2)\\ \text{or}\\ (a_1,a_2)\to(a_1^*,a_2^*)}} I\left(a_1^*, a_2^*\right) = 0. \tag{9.2.15}$$

Case now of $0 < \alpha_1, \alpha_2 < 1$, then

$$I\left(a_1^*, a_2^*\right) =$$

$$\int_{a_1^*}^{b_1}\int_{a_2^*}^{b_2}\left(\left(t_1 - a_1^*\right)^{\alpha_1-1}\left(t_2 - a_2^*\right)^{\alpha_2-1} - (t_1 - a_1)^{\alpha_1-1}(t_2 - a_2)^{\alpha_2-1}\right)dt_1 dt_2$$

$$= \frac{\left(b_1 - a_1^*\right)^{\alpha_1}}{\alpha_1}\frac{\left(b_2 - a_2^*\right)^{\alpha_2}}{\alpha_2} -$$

$$\left(\frac{(b_1 - a_1)^{\alpha_1} - \left(a_1^* - a_1\right)^{\alpha_1}}{\alpha_1}\right)\left(\frac{(b_2 - a_2)^{\alpha_2} - \left(a_2^* - a_2\right)^{\alpha_2}}{\alpha_2}\right). \tag{9.2.16}$$

Hence, when $0 < \alpha_1, \alpha_2 < 1$, we get

$$\lim_{\substack{(a_1^*,a_2^*)\to(a_1,a_2)\\ (a_1,a_2)\to(a_1^*,a_2^*)}} I\left(a_1^*, a_2^*\right) = 0. \tag{9.2.17}$$

We observe that

$$I\left(a_1^*, a_2^*\right) \leq I^*\left(a_1^*, a_2^*\right) :=$$

$$\int_{a_1^*}^{b_1}\int_{a_2^*}^{b_2}(t_1 - a_1)^{\alpha_1-1}\left|(t_2 - a_2)^{\alpha_2-1} - \left(t_2 - a_2^*\right)^{\alpha_2-1}\right| dt_1 dt_2$$

$$+\int_{a_1^*}^{b_1}\int_{a_2^*}^{b_2}\left(t_2 - a_2^*\right)^{\alpha_2-1}\left|(t_1 - a_1)^{\alpha_1-1} - \left(t_1 - a_1^*\right)^{\alpha_1-1}\right| dt_1 dt_2. \tag{9.2.18}$$

Next we treat the case of $\alpha_1 > 1, 0 < \alpha_2 < 1$.

Therefore it holds

$$I^*\left(a_1^*, a_2^*\right) = \int_{a_1^*}^{b_1}\int_{a_2^*}^{b_2}(t_1 - a_1)^{\alpha_1-1}\left(\left(t_2 - a_2^*\right)^{\alpha_2-1} - (t_2 - a_2)^{\alpha_2-1}\right)dt_1 dt_2 \tag{9.2.19}$$

$$+\int_{a_1^*}^{b_1}\int_{a_2^*}^{b_2}\left(t_2 - a_2^*\right)^{\alpha_2-1}\left((t_1 - a_1)^{\alpha_1-1} - \left(t_1 - a_1^*\right)^{\alpha_1-1}\right)dt_1 dt_2 =$$

$$\left(\frac{(b_1-a_1)^{\alpha_1}-(a_1^*-a_1)^{\alpha_1}}{\alpha_1}\right)\left(\frac{(b_2-a_2^*)^{\alpha_2}}{\alpha_2}-\frac{(b_2-a_2)^{\alpha_2}}{\alpha_2}+\frac{(a_2^*-a_2)^{\alpha_2}}{\alpha_2}\right)$$

$$+\frac{(b_2-a_2^*)^{\alpha_2}}{\alpha_2}\left(\frac{(b_1-a_1)^{\alpha_1}}{\alpha_1}-\frac{(a_1^*-a_1)^{\alpha_1}}{\alpha_1}-\frac{(b_1-a_1^*)^{\alpha_1}}{\alpha_1}\right).$$

Clearly then ($\alpha_1 > 1, 0 < \alpha_2 < 1$)

$$\lim_{\substack{(a_1,a_2)\to(a_1^*,a_2^*)\\ \text{or}\\ (a_1^*,a_2^*)\to(a_1,a_2)}} I\left(a_1^*, a_2^*\right) = 0. \tag{9.2.20}$$

Finally, we prove the case of $\alpha_2 > 1$ and $0 < \alpha_1 < 1$. We have that

$$I^*\left(a_1^*, a_2^*\right) = \int_{a_1^*}^{b_1}\int_{a_2^*}^{b_2}(t_1-a_1)^{\alpha_1-1}\left((t_2-a_2)^{\alpha_2-1}-\left(t_2-a_2^*\right)^{\alpha_2-1}\right)dt_1dt_2$$

$$+\int_{a_1^*}^{b_1}\int_{a_2^*}^{b_2}\left(t_2-a_2^*\right)^{\alpha_2-1}\left(\left(t_1-a_1^*\right)^{\alpha_1-1}-(t_1-a_1)^{\alpha_1-1}\right)dt_1dt_2 = \tag{9.2.21}$$

$$\left(\frac{(b_1-a_1)^{\alpha_1}-(a_1^*-a_1)^{\alpha_1}}{\alpha_1}\right)\left(-\frac{(b_2-a_2^*)^{\alpha_2}}{\alpha_2}+\frac{(b_2-a_2)^{\alpha_2}}{\alpha_2}-\frac{(a_2^*-a_2)^{\alpha_2}}{\alpha_2}\right)$$

$$+\frac{(b_2-a_2^*)^{\alpha_2}}{\alpha_2}\left(-\frac{(b_1-a_1)^{\alpha_1}}{\alpha_1}+\frac{(a_1^*-a_1)^{\alpha_1}}{\alpha_1}+\frac{(b_1-a_1^*)^{\alpha_1}}{\alpha_1}\right).$$

Clearly then ($\alpha_2 > 1, 0 < \alpha_1 < 1$)

$$\lim_{\substack{(a_1,a_2)\to(a_1^*,a_2^*)\\ \text{or}\\ (a_1^*,a_2^*)\to(a_1,a_2)}} I\left(a_1^*, a_2^*\right) = 0. \tag{9.2.22}$$

We proved $\rho = 0$, and $\delta = 0$ in all cases of this section.

The case of $a_1 > a_1^*$ and $a_2 > a_2^*$, as symmetric to the already treated one of $a_1 < a_1^*$ and $a_2 < a_2^*$, is omitted.

(II) The remaining cases are: let $a_1, a_1^*, b_1 \in [a, b]$; $a_2, a_2^*, b_2 \in [c, d]$, we can have

$$\begin{gathered}(\mathrm{II}_1)\ a_1 > a_1^* \text{ and } a_2 < a_2^*,\\ \text{or}\\ (\mathrm{II}_2)\ a_1 < a_1^* \text{ and } a_2 > a_2^*.\end{gathered} \tag{9.2.23}$$

Notice that the subcases (II_1) and (II_2) are symmetric, and treated the same way. As such we treat only the case (II_2).

We observe again that

$$F(a_1, a_2) - F(a_1^*, a_2^*) = \tag{9.2.24}$$

$$\int_{a_1}^{b_1}\int_{a_2}^{b_2}(t_1 - a_1)^{\alpha_1 - 1}(t_2 - a_2)^{\alpha_2 - 1} f(t_1, t_2)\,dt_1 dt_2 -$$

$$\int_{a_1^*}^{b_1}\int_{a_2^*}^{b_2}(t_1 - a_1^*)^{\alpha_1 - 1}(t_2 - a_2^*)^{\alpha_2 - 1} f(t_1, t_2)\,dt_1 dt_2 =$$

$$\int_{a_1}^{a_1^*}\int_{a_2}^{b_2}(t_1 - a_1)^{\alpha_1 - 1}(t_2 - a_2)^{\alpha_2 - 1} f(t_1, t_2)\,dt_1 dt_2 +$$

$$\int_{a_1^*}^{b_1}\int_{a_2}^{b_2}(t_1 - a_1)^{\alpha_1 - 1}(t_2 - a_2)^{\alpha_2 - 1} f(t_1, t_2)\,dt_1 dt_2 -$$

$$\int_{a_1^*}^{b_1}\int_{a_2^*}^{a_2}(t_1 - a_1^*)^{\alpha_1 - 1}(t_2 - a_2^*)^{\alpha_2 - 1} f(t_1, t_2)\,dt_1 dt_2 - \tag{9.2.25}$$

$$\int_{a_1^*}^{b_1}\int_{a_2}^{b_2}(t_1 - a_1^*)^{\alpha_1 - 1}(t_2 - a_2^*)^{\alpha_2 - 1} f(t_1, t_2)\,dt_1 dt_2 =$$

$$\int_{a_1^*}^{b_1}\int_{a_2}^{b_2}\left((t_1 - a_1)^{\alpha_1 - 1}(t_2 - a_2)^{\alpha_2 - 1} - (t_1 - a_1^*)^{\alpha_1 - 1}(t_2 - a_2^*)^{\alpha_2 - 1}\right) f(t_1, t_2)\,dt_1 dt_2$$

$$+ \int_{a_1}^{a_1^*}\int_{a_2}^{b_2}(t_1 - a_1)^{\alpha_1 - 1}(t_2 - a_2)^{\alpha_2 - 1} f(t_1, t_2)\,dt_1 dt_2 - \tag{9.2.26}$$

$$\int_{a_1^*}^{b_1}\int_{a_2^*}^{a_2}(t_1 - a_1^*)^{\alpha_1 - 1}(t_2 - a_2^*)^{\alpha_2 - 1} f(t_1, t_2)\,dt_1 dt_2.$$

Call

$$I(a_1^*, a_2) :=$$

$$\int_{a_1^*}^{b_1}\int_{a_2}^{b_2}\left|(t_1 - a_1)^{\alpha_1 - 1}(t_2 - a_2)^{\alpha_2 - 1} - (t_1 - a_1^*)^{\alpha_1 - 1}(t_2 - a_2^*)^{\alpha_2 - 1}\right| dt_1 dt_2. \tag{9.2.27}$$

Hence, we have

$$\left|F\left(a_1,a_2\right)-F\left(a_1^*,a_2^*\right)\right|\le$$

$$\left\{I\left(a_1^*,a_2\right)+\frac{\left(a_1^*-a_1\right)^{\alpha_1}}{\alpha_1}\frac{\left(b_2-a_2\right)^{\alpha_2}}{\alpha_2}+\frac{\left(b_1-a_1^*\right)^{\alpha_1}}{\alpha_1}\frac{\left(a_2-a_2^*\right)^{\alpha_2}}{\alpha_2}\right\}\|f\|_\infty. \tag{9.2.28}$$

Therefore it holds

$$\delta:=\lim_{\substack{|a_1-a_1^*|\to0,\\|a_2-a_2^*|\to0}}\left|F\left(a_1,a_2\right)-F\left(a_1^*,a_2^*\right)\right|\le\left(\lim_{\substack{|a_1-a_1^*|\to0,\\|a_2-a_2^*|\to0}}I\left(a_1^*,a_2\right)\right)\|f\|_\infty=:\theta. \tag{9.2.29}$$

We will prove that $\theta=0$, hence $\delta=0$, in all possible cases.

If $\alpha_1=\alpha_2=1$, then $I\left(a_1^*,a_2\right)=0$, hence $\theta=0$.

If $\alpha_1=1$, $\alpha_2>0$ we get

$$I\left(a_1^*,a_2\right)=\left(b_1-a_1^*\right)\int_{a_2}^{b_2}\left|\left(t_2-a_2\right)^{\alpha_2-1}-\left(t_2-a_2^*\right)^{\alpha_2-1}\right|dt_2. \tag{9.2.30}$$

Assume $\alpha_2>1$, then $\alpha_2-1>0$. Hence

$$I\left(a_1^*,a_2\right)=\left(b_1-a_1^*\right)\int_{a_2}^{b_2}\left(\left(t_2-a_2^*\right)^{\alpha_2-1}-\left(t_2-a_2\right)^{\alpha_2-1}\right)dt_2$$

$$=\left(b_1-a_1^*\right)\left\{\frac{\left(b_2-a_2^*\right)^{\alpha_2}}{\alpha_2}-\frac{\left(a_2-a_2^*\right)^{\alpha_2}}{\alpha_2}-\frac{\left(b_2-a_2\right)^{\alpha_2}}{\alpha_2}\right\}. \tag{9.2.31}$$

Clearly, then

$$\lim_{|a_2-a_2^*|\to0,}I\left(a_1^*,a_2\right)=0, \tag{9.2.32}$$

hence $\theta=0$.

Let the case now of $\alpha_2=1$, $\alpha_1>1$. Then

$$I\left(a_1^*,a_2\right)=\left(b_2-a_2\right)\int_{a_1^*}^{b_1}\left|\left(t_1-a_1\right)^{\alpha_1-1}-\left(t_1-a_1^*\right)^{\alpha_1-1}\right|dt_1$$

$$=\left(b_2-a_2\right)\int_{a_1^*}^{b_1}\left(\left(t_1-a_1\right)^{\alpha_1-1}-\left(t_1-a_1^*\right)^{\alpha_1-1}\right)dt_1 \tag{9.2.33}$$

$$=\left(b_2-a_2\right)\left(\frac{\left(b_1-a_1\right)^{\alpha_1}}{\alpha_1}-\frac{\left(a_1^*-a_1\right)^{\alpha_1}}{\alpha_1}-\frac{\left(b_1-a_1^*\right)^{\alpha_1}}{\alpha_1}\right).$$

Then $\theta=0$.

If $\alpha_1 = 1$, and $0 < \alpha_2 < 1$, then $\alpha_2 - 1 < 0$. Hence

$$I\left(a_1^*, a_2\right) = \left(b_1 - a_1^*\right) \int_{a_2}^{b_2} \left((t_2 - a_2)^{\alpha_2 - 1} - \left(t_2 - a_2^*\right)^{\alpha_2 - 1}\right) dt_2 =$$

$$\left(b_1 - a_1^*\right) \left\{ \frac{(b_2 - a_2)^{\alpha_2}}{\alpha_2} - \frac{\left(b_2 - a_2^*\right)^{\alpha_2}}{\alpha_2} + \frac{\left(a_2 - a_2^*\right)^{\alpha_2}}{\alpha_2} \right\}, \quad (9.2.34)$$

hence $\theta = 0$.

Let now $\alpha_2 = 1$, $0 < \alpha_1 < 1$. Then

$$I\left(a_1^*, a_2\right) = (b_2 - a_2) \int_{a_1^*}^{b_1} \left(\left(t_1 - a_1^*\right)^{\alpha_1 - 1} - (t_1 - a_1)^{\alpha_1 - 1}\right) dt_1$$

$$= (b_2 - a_2) \left\{ \frac{\left(b_1 - a_1^*\right)^{\alpha_1}}{\alpha_1} - \frac{(b_1 - a_1)^{\alpha_1}}{\alpha_1} + \frac{\left(a_1^* - a_1\right)^{\alpha_1}}{\alpha_1} \right\}, \quad (9.2.35)$$

hence $\theta = 0$.

We observe that:

$$I\left(a_1^*, a_2\right) \leq \int_{a_1^*}^{b_1} \int_{a_2}^{b_2} (t_1 - a_1)^{\alpha_1 - 1} \left|(t_2 - a_2)^{\alpha_2 - 1} - \left(t_2 - a_2^*\right)^{\alpha_2 - 1}\right| dt_1 dt_2$$

$$+ \int_{a_1^*}^{b_1} \int_{a_2}^{b_2} \left(t_2 - a_2^*\right)^{\alpha_2 - 1} \left|(t_1 - a_1)^{\alpha_1 - 1} - \left(t_1 - a_1^*\right)^{\alpha_1 - 1}\right| dt_1 dt_2 =: J\left(a_1^*, a_2\right). \quad (9.2.36)$$

I.e.

$$I\left(a_1^*, a_2\right) \leq J\left(a_1^*, a_2\right). \quad (9.2.37)$$

Case of $\alpha_1, \alpha_2 > 1$. Then

$$J\left(a_1^*, a_2\right) = \int_{a_1^*}^{b_1} \int_{a_2}^{b_2} (t_1 - a_1)^{\alpha_1 - 1} \left(\left(t_2 - a_2^*\right)^{\alpha_2 - 1} - (t_2 - a_2)^{\alpha_2 - 1}\right) dt_1 dt_2$$

$$+ \int_{a_1^*}^{b_1} \int_{a_2}^{b_2} \left(t_2 - a_2^*\right)^{\alpha_2 - 1} \left((t_1 - a_1)^{\alpha_1 - 1} - \left(t_1 - a_1^*\right)^{\alpha_1 - 1}\right) dt_1 dt_2 = \quad (9.2.38)$$

$$\left(\frac{(b_1 - a_1)^{\alpha_1}}{\alpha_1} - \frac{\left(a_1^* - a_1\right)^{\alpha_1}}{\alpha_1}\right) \left\{ \frac{\left(b_2 - a_2^*\right)^{\alpha_2}}{\alpha_2} - \frac{\left(a_2 - a_2^*\right)^{\alpha_2}}{\alpha_2} - \frac{(b_2 - a_2)^{\alpha_2}}{\alpha_2} \right\}$$

$$+\left(\frac{\left(b_2-a_2^*\right)^{\alpha_2}}{\alpha_2}-\frac{\left(a_2-a_2^*\right)^{\alpha_2}}{\alpha_2}\right)\left\{\frac{(b_1-a_1)^{\alpha_1}}{\alpha_1}-\frac{\left(a_1^*-a_1\right)^{\alpha_1}}{\alpha_1}-\frac{\left(b_1-a_1^*\right)^{\alpha_1}}{\alpha_1}\right\}, \tag{9.2.39}$$

hence $\theta = 0$.

Case of $0 < \alpha_1, \alpha_2 < 1$, then

$$J\left(a_1^*, a_2\right) = \int_{a_1^*}^{b_1}\int_{a_2}^{b_2}(t_1-a_1)^{\alpha_1-1}\left((t_2-a_2)^{\alpha_2-1}-\left(t_2-a_2^*\right)^{\alpha_2-1}\right)dt_1dt_2$$

$$+\int_{a_1^*}^{b_1}\int_{a_2}^{b_2}\left(t_2-a_2^*\right)^{\alpha_2-1}\left(\left(t_1-a_1^*\right)^{\alpha_1-1}-(t_1-a_1)^{\alpha_1-1}\right)dt_1dt_2 = \tag{9.2.40}$$

$$\left(\frac{(b_1-a_1)^{\alpha_1}}{\alpha_1}-\frac{\left(a_1^*-a_1\right)^{\alpha_1}}{\alpha_1}\right)\left\{\frac{(b_2-a_2)^{\alpha_2}}{\alpha_2}-\frac{\left(b_2-a_2^*\right)^{\alpha_2}}{\alpha_2}+\frac{\left(a_2-a_2^*\right)^{\alpha_2}}{\alpha_2}\right\}$$

$$+\left(\frac{\left(b_2-a_2^*\right)^{\alpha_2}}{\alpha_2}-\frac{\left(a_2-a_2^*\right)^{\alpha_2}}{\alpha_2}\right)\left\{\frac{\left(b_1-a_1^*\right)^{\alpha_1}}{\alpha_1}-\frac{(b_1-a_1)^{\alpha_1}}{\alpha_1}+\frac{\left(a_1^*-a_1\right)^{\alpha_1}}{\alpha_1}\right\}, \tag{9.2.41}$$

hence $\theta = 0$.

Next case of $\alpha_1 > 1, 0 < \alpha_2 < 1$. We observe that

$$J\left(a_1^*, a_2\right) = \int_{a_1^*}^{b_1}\int_{a_2}^{b_2}(t_1-a_1)^{\alpha_1-1}\left((t_2-a_2)^{\alpha_2-1}-\left(t_2-a_2^*\right)^{\alpha_2-1}\right)dt_1dt_2 \tag{9.2.42}$$

$$+\int_{a_1^*}^{b_1}\int_{a_2}^{b_2}\left(t_2-a_2^*\right)^{\alpha_2-1}\left((t_1-a_1)^{\alpha_1-1}-\left(t_1-a_1^*\right)^{\alpha_1-1}\right)dt_1dt_2 =$$

$$\left(\frac{(b_1-a_1)^{\alpha_1}}{\alpha_1}-\frac{\left(a_1^*-a_1\right)^{\alpha_1}}{\alpha_1}\right)\left\{\frac{(b_2-a_2)^{\alpha_2}}{\alpha_2}-\frac{\left(b_2-a_2^*\right)^{\alpha_2}}{\alpha_2}+\frac{\left(a_2-a_2^*\right)^{\alpha_2}}{\alpha_2}\right\} \tag{9.2.43}$$

$$+\left(\frac{\left(b_2-a_2^*\right)^{\alpha_2}}{\alpha_2}-\frac{\left(a_2-a_2^*\right)^{\alpha_2}}{\alpha_2}\right)\left\{\frac{(b_1-a_1)^{\alpha_1}}{\alpha_1}-\frac{\left(a_1^*-a_1\right)^{\alpha_1}}{\alpha_1}-\frac{\left(b_1-a_1^*\right)^{\alpha_1}}{\alpha_1}\right\},$$

hence $\theta = 0$.

Finally, we prove the case of $\alpha_2 > 1$ and $0 < \alpha_1 < 1$. In that case it holds

$$J\left(a_1^*, a_2\right) = \int_{a_1^*}^{b_1}\int_{a_2}^{b_2}(t_1-a_1)^{\alpha_1-1}\left(\left(t_2-a_2^*\right)^{\alpha_2-1}-(t_2-a_2)^{\alpha_2-1}\right)dt_1dt_2$$

$$+\int_{a_1^*}^{b_1}\int_{a_2}^{b_2}\left(t_2-a_2^*\right)^{\alpha_2-1}\left(\left(t_1-a_1^*\right)^{\alpha_1-1}-\left(t_1-a_1\right)^{\alpha_1-1}\right)dt_1dt_2= \tag{9.2.44}$$

$$\left(\frac{(b_1-a_1)^{\alpha_1}-\left(a_1^*-a_1\right)^{\alpha_1}}{\alpha_1}\right)\left\{\frac{\left(b_2-a_2^*\right)^{\alpha_2}}{\alpha_2}-\frac{\left(a_2-a_2^*\right)^{\alpha_2}}{\alpha_2}-\frac{(b_2-a_2)^{\alpha_2}}{\alpha_2}\right\}$$

$$+\left(\frac{\left(b_2-a_2^*\right)^{\alpha_2}-\left(a_2-a_2^*\right)^{\alpha_2}}{\alpha_2}\right)\left\{\frac{\left(b_1-a_1^*\right)^{\alpha_1}}{\alpha_1}-\frac{(b_1-a_1)^{\alpha_1}}{\alpha_1}+\frac{\left(a_1^*-a_1\right)^{\alpha_1}}{\alpha_1}\right\}, \tag{9.2.45}$$

hence $\theta=0$.

We have proved that $\delta=0$, in all possible subcases of (II_2).

We have proved that F is a continuous function over $[a,b_1]\times[c,b_2]$. ■

Now we can state:

Theorem 9.3 *Let $f\in L_\infty\left(\prod_{i=1}^k[a_i,b_i]\right)$, $\alpha_i>0$, $i=1,...,k\in\mathbb{N}$. Consider the function*

$$F(x_1,...,x_k)=\int_{x_1}^{b_1^*}...\int_{x_k}^{b_k^*}\prod_{i=1}^k(t_i-x_i)^{\alpha_i-1}f(t_1,...,t_k)\,dt_1...dt_k, \tag{9.2.46}$$

where $a_i\le x_i\le b_i^*\le b_i$, $i=1,...,k$.

Then F is continuous on $\prod_{i=1}^k\left[a_i,b_i^*\right]$.

Remark 9.4 In the setting of Theorem 9.3: Consider the right multidimensional Riemann–Liouville fractional integral of order $\alpha=(\alpha_1,...,\alpha_k)$, $\alpha_i>0$, $i=1,...,k$:

$$\left(I_{b^*-}^{\alpha}f\right)(x)=\frac{1}{\prod_{i=1}^k\Gamma(\alpha_i)}\int_{x_1}^{b_1^*}...\int_{x_k}^{b_k^*}\prod_{i=1}^k(t_i-x_i)^{\alpha_i-1}f(t_1,...,t_k)\,dt_1...dt_k, \tag{9.2.47}$$

where $a_i\le x_i\le b_i^*\le b_i$, $i=1,...,k$, where $b^*=\left(b_1^*,...,b_k^*\right)$, $x=(x_1,...,x_k)$, Γ is the gamma function.

By Theorem 9.3 we get that $\left(I_{b^*-}^{\alpha}f\right)$ is a continuous function for every $x\in\prod_{i=1}^k\left[a_i,b_i^*\right]$.

We notice that

$$\left|\left(I_{b_-^*}^{\alpha} f\right)(x)\right| \leq \frac{1}{\prod_{i=1}^{k} \Gamma(\alpha_i)} \left(\int_{x_1}^{b_1^*} \ldots \int_{x_k}^{b_k^*} \prod_{i=1}^{k} (t_i - x_i)^{\alpha_i - 1} dt_1 \ldots dt_k\right) \|f\|_\infty \tag{9.2.48}$$

$$= \frac{1}{\prod_{i=1}^{k} \Gamma(\alpha_i)} \prod_{i=1}^{k} \left(\int_{x_i}^{b_i^*} (t_i - x_i)^{\alpha_i - 1} dt_i\right) \|f\|_\infty =$$

$$\frac{1}{\prod_{i=1}^{k} \Gamma(\alpha_i)} \prod_{i=1}^{k} \frac{\left(b_i^* - x_i\right)^{\alpha_i}}{\alpha_i} \|f\|_\infty = \left(\prod_{i=1}^{k} \frac{\left(b_i^* - x_i\right)^{\alpha_i}}{\Gamma(\alpha_i + 1)}\right) \|f\|_\infty. \tag{9.2.49}$$

That is

$$\left|\left(I_{b_-^*}^{\alpha} f\right)(x)\right| \leq \left(\prod_{i=1}^{k} \frac{\left(b_i^* - x_i\right)^{\alpha_i}}{\Gamma(\alpha_i + 1)}\right) \|f\|_\infty. \tag{9.2.50}$$

In particular we get

$$\left(I_{b_-^*}^{\alpha} f\right)(b^*) = 0, \tag{9.2.51}$$

and

$$\left\|I_{b_-^*}^{\alpha} f\right\|_{\infty, \prod_{i=1}^{k} [a_i, b_i^*]} \leq \left(\prod_{i=1}^{k} \frac{\left(b_i^* - a_i\right)^{\alpha_i}}{\Gamma(\alpha_i + 1)}\right) \|f\|_\infty. \tag{9.2.52}$$

That is $I_{b_-^*}^{\alpha} f$ is a bounded linear operator, which here is also a positive operator.

Reference

1. G.A. Anastassiou, Fractional representation formulae and right fractional inequalities. Math. Comput. Model. **54**, 3098–3115 (2011)

Printed in the United States
By Bookmasters

Printed in the United States
By Bookmasters